The Urbana Free Library

To renew: call 217-367-4057
or go to "*urbanafreelibrary.org*"
and select "Renew/Request Items"

Umami

ARTS AND TRADITIONS OF THE TABLE:
PERSPECTIVES ON CULINARY HISTORY

Arts and Traditions of the Table: Perspectives on Culinary History

Albert Sonnenfeld, Series Editor

Salt: Grain of Life, Pierre Laszlo, translated by Mary Beth Mader

Culture of the Fork, Giovanni Rebora, translated by Albert Sonnenfeld

French Gastronomy: The History and Geography of a Passion, Jean-Robert Pitte, translated by Jody Gladding

Pasta: The Story of a Universal Food, Silvano Serventi and Françoise Sabban, translated by Antony Shugar

Slow Food: The Case for Taste, Carlo Petrini, translated by William McCuaig

Italian Cuisine: A Cultural History, Alberto Capatti and Massimo Montanari, translated by Áine O'Healy

British Food: An Extraordinary Thousand Years of History, Colin Spencer

A Revolution in Eating: How the Quest for Food Shaped America, James E. McWilliams

Sacred Cow, Mad Cow: A History of Food Fears, Madeleine Ferrières, translated by Jody Gladding

Molecular Gastronomy: Exploring the Science of Flavor, Hervé This, translated by M. B. DeBevoise

Food Is Culture, Massimo Montanari, translated by Albert Sonnenfeld

Kitchen Mysteries: Revealing the Science of Cooking, Hervé This, translated by Jody Gladding

Hog and Hominy: Soul Food from Africa to America, Frederick Douglass Opie

Gastropolis: Food and New York City, edited by Annie Hauck-Lawson and Jonathan Deutsch

Building a Meal: From Molecular Gastronomy to Culinary Constructivism, Hervé This, translated by M. B. DeBevoise

Eating History: Thirty Turning Points in the Making of American Cuisine, Andrew F. Smith

The Science of the Oven, Hervé This, translated by Jody Gladding

Pomodoro! A History of the Tomato in Italy, David Gentilcore

Cheese, Pears, and History in a Proverb, Massimo Montanari, translated by Beth Archer Brombert

Food and Faith in Christian Culture, edited by Ken Albala and Trudy Eden

The Kitchen as Laboratory: Reflections on the Science of Food and Cooking, edited by César Vega, Job Ubbink, and Erik van der Linden

Creamy and Crunchy: An Informal History of Peanut Butter, the All-American Food, Jon Krampner

Let the Meatballs Rest: And Other Stories About Food and Culture, Massimo Montanari, translated by Beth Archer Brombert

The Secret Financial Life of Food: From Commodities Markets to Supermarkets, Kara Newman

Drinking History: Fifteen Turning Points in the Making of American Beverages, Andrew Smith

Italian Identity in the Kitchen, or Food and the Nation, Massimo Montanari, translated by Beth Archer Brombert

Fashioning Appetite: Restaurants and the Making of Modern Identity, Joanne Finkelstein

The Land of the Five Flavors: A Cultural History of Chinese Cuisine, Thomas O. Höllmann, translated by Karen Margolis

The Insect Cookbook: Food for a Sustainable Planet, Arnold van Huis, Henk van Gurp, and Marcel Dicke, translated by Françoise Takken-Kaminker and Diane Blumenfeld-Schaap

Religion, Food, and Eating in North America, edited by Benjamin E. Zeller, Marie W. Dallam, Reid L. Neilson, and Nora L. Rubel

Ole G. Mouritsen and Klavs Styrbæk

Umami
Unlocking the Secrets of the Fifth Taste

Photography, layout, and design
Jonas Drotner Mouritsen

Translation and adaptation to English
Mariela Johansen

COLUMBIA UNIVERSITY PRESS
NEW YORK

Columbia University Press
Publishers Since 1893
New York   Chichester, West Sussex
cup.columbia.edu
Copyright © 2014 Columbia University Press
All rights reserved

Library of Congress Cataloging-in-Publication Data

Mouritsen, Ole G.
  Umami: unlocking the secrets of the fifth taste / Ole G. Mouritsen and Klavs Styrbæk
    p. cm. — (Arts and traditions of the table: perspectives on culinary history)
  Includes index
  ISBN 978-0-231-16890-8 (cloth : alk. paper) — ISBN 978-0-231-53758-2 (e-book)

Library of Congress Holding Information can be found on the Library of Congress Online Catalog.

2013952514

∞

Columbia University Press books are printed on permanent and durable acid-free paper.
This book is printed on paper with recycled content.
Printed in the United States of America

c 10 9 8 7 6 5 4 3 2 1

Cover design by Jonas Drotner Mouritsen.

www.umamibook.net

References to websites (URLs) were accurate at the time of writing. Neither the author nor Columbia University Press is responsible for URLs that may have expired or changed since the manuscript was prepared.

Contents

ACKNOWLEDGMENTS ix

PROLOGUE: HOW IT ALL BEGAN xiii

WHAT EXACTLY IS TASTE, AND WHY IS IT IMPORTANT? 1
The basic tastes: From seven to four to five and possibly many more 1
Why do we need to be able to taste our food? 4
There is more to it: Sensory science, taste, smell, aroma, flavor, mouthfeel, texture, and chemesthesis 5
Is there a taste map of the tongue? 7
Why are some foods more palatable than others? 8
A few words about proteins, amino acids, nucleotides, nucleic acids, and enzymes 9
Glutamic acid, glutamate, and the glutamate ion 11
Glutamic acid and glutamate in our food 12
How does glutamate taste, and how little is required for us to taste it? 13

THE FIRST FOUR: SOUR, SWEET, SALTY, AND BITTER 15
The physiology and biochemistry of taste 15
The interplay between sweet and bitter 16
Taste receptors: This is how they work 17
When words fail us: Descriptions of tastes 20

THE FIFTH TASTE: WHAT IS UMAMI? 23
Science, soup, and the search for the fifth taste 23
Glutamic acid and glutamate 24
What is the meaning of the word *umami*? 26
From laboratory to mass production 27
How MSG is made 28
A little letter with a huge impact: The 'Chinese restaurant syndrome' 32
The Japanese discover other umami substances 34
It all starts with mother's milk 35

Umami as a global presence 36
Umami has won acceptance as a distinct taste 38
And umami is still controversial … 39

1 + 1 = 8: GUSTATORY SYNERGY 41
Amazing interplay: Basal and synergistic umami 41
Detecting umami synergy on the tongue and in the brain 42
Japanese dashi: *The* textbook example of umami synergy 43
The art of making Japanese dashi 45
Nordic dashi 47
Dashi closer to home—a Japanese soup with a Scandinavian twist 48
Seaweeds enhance the umami in fish 52
How to make smoked shrimp heads 53
Many substances interact synergistically with umami 54
A breakthrough discovery of yet another synergistic substance 54
The interplay between glutamate and the four classic tastes 55
A simple taste test: Umami vs. salt 56
Umami-rich 'foie gras from the sea' 57
Food pairing and umami 60
Creating tastes synthetically 60
Umami: Either as little or as much as you like 62

UMAMI FROM THE OCEANS: SEAWEEDS, FISH, AND SHELLFISH 65
Seaweeds and konbu: *The* mother lode of umami 65
A world of konbu in Japan 66
Fresh fish and shellfish 69
Cooked fish and shellfish dishes and soups 69
Umami and the art of killing a fish 72
A traditional clambake: New England method, Danish ingredients 74
Everyday umami in ancient Greece and Rome 79

Fish sauces and fish pastes 81
Modern *garum* 85
Shellfish paste 87
Oyster sauce 87
Sushi and fermented fish 88
Katsuobushi 90
Catching *katsuo* to optimize umami 91
Niboshi 91
The hardest foodstuff in the world 92
Kusaya 96
Nordic variations: Horrible smells and heavenly tastes 96
Fish roe 98
Seven friends, *The Compleat Angler,* and a pike 100

UMAMI FROM THE LAND: FUNGI AND PLANTS 105
Umami from the plant kingdom 105
Dried fungi 110
Fermented soybeans 111
Soy sauce 112
Production of *shōyu* 113
Miso 114
Production of miso 114
The Asian answer to cheese: Fermented soybean cakes 118
Nattō 120
Black garlic 122
Shōjin ryōri: An old tradition with a modern presence 122
The enlightened kitchen 124
Tomatoes 126
Green tea 134

UMAMI FROM LAND ANIMALS: MEAT, EGGS, AND DAIRY PRODUCTS 137
The animal kingdom delivers umami in spades 137
Homo sapiens is a cook 140
Preserving meats in the traditional ways 142
Air-dried hams 143
Salted beef: Pastrami and corned beef 144
Bacon and sausages 145
Dairy products 146
Blue cheeses 146
Aged, dried, and hard cheeses 148
Eggs and mayonnaise 151
Harry's crème from Harry's Bar 151

UMAMI: THE SECRET BEHIND THE HUMBLE SOUP STOCK 155
Soup is umami 155
Osmazome and *The Physiology of Taste* 158
Amino acids in soup stocks 160
A real find: A dashi bar 160
The taste of a beef stock 162
Ready-made umami 164
Knorr and Maggi: European umami pioneers 165

MAKING THE MOST OF UMAMI 167
MSG as a food additive 167
Other commercial sources of umami 168
Hydrolyzed protein 169
Umami in a jar 170
Yeast extract 172
Nutritional yeast 172
More sources of umami for vegans 173
Ketchup 174
Bagna càuda 175
Worcestershire sauce 176
Umami in a tube 177
Twelve easy ways to add umami 178
Quintessentially Danish: Brown gravy, *medisterpølse*, and beef patties 180
Slow cooking: The secret of more umami 182
Ratatouille and *brandade* 190
This is why fast food tastes so good 191
Green salads and raw vegetables 194
Umami in dishes made with small fowl 196
Cooked potatoes: Nothing could be simpler 197
Rice and sake 197
Beer 200
Umami in sweets 202
Mirin is a sweet rice wine with umami 203

UMAMI AND WELLNESS 207

Umami and MSG: Food without 'chemicals' 207
Umami satisfies the appetite 209
Why does umami make us feel full?
   The 'brain' in the stomach 209
Umami for a sick and aging population 210
Umami for life 211

EPILOGUE: UMAMI HAS COME TO STAY 213

TECHNICAL AND SCIENTIFIC DETAILS 217

Umami and the first glutamate receptor 217
Yet another receptor for umami 218

Umami synergy 220
The taste of amino acids 222
Taste thresholds for umami 223
Content of glutamate and 5'-ribonucleotides
   in different foods 223

BIBLIOGRAPHY 233
ILLUSTRATION CREDITS 237
GLOSSARY 239
INDEX 255

THE PEOPLE BEHIND THE BOOK 264

RECIPES

Potato water dashi with smoked shrimp heads 53
Monkfish liver au gratin with
   crabmeat and vegetables 58
Pearled spelt, beets, and lobster 70
Crab soup 76
Clambake in a pot 78
Patina de pisciculis 82
Garum 86
Quick-and-easy *garum* 86
Smoked quick-and-easy *garum* 87
Seriously old-fashioned sourdough rye bread 107
Anchovies, grilled onions, sourdough bread,
   pata negra ham, and mushrooms 108
Deep-fried eggplants with miso (*nasu dengaku*) 115
White asparagus in miso with oysters,
   cucumber oil, and small fish 116
Grilled *shōjin kabayaki:* 'fried eel'
   made from lotus root 123
Baked monkfish liver with raspberries
   and peanuts 128
Slow-roasted sauce with tomatoes,
   root vegetables, and herbs 130
Fried mullet with baked grape tomatoes,
   marinated sago pearls, and black garlic 132
Mushrooms, foie gras, and mushroom essence 138

Parmesan biscuits with bacon and yeast flakes 150
Harry's crème 152
Chicken bouillon 157
Green pea soup with scallops and seaweed 163
Dressing with nutritional yeast 173
Eggplant gratinée with garlic, anchovies,
   and nutritional yeast 174
Oysters au gratin with a crust of nutritional
   yeast and smoked shrimp head powder 175
Bagna càuda 176
Old-fashioned Danish *medisterpølse* 181
Beef patties, Danish style 183
Chicken Marengo 185
Cassoulet 186
Beef *estofado* 188
Sicilian ratatouille 190
Brandade with air-dried ham and green peas 191
Three-day pizza with umami—not
   really a 'fast food' 192
Quail pâté 196
Risotto 197
Oxtails braised in wheat beer 201
Umami sorbet with *maccha* and tomato 202
White chocolate cream, black sesame seeds,
   Roquefort, and brioche with nutritional yeast 203

Acknowledgments

The undertaking of a joint project that encompasses as many diverse aspects of a topic as this book does is rarely possible without the assistance and support of a wide range of individuals and organizations. In the course of the many months that went into gathering the material, testing recipes in laboratories and kitchens, and exploring new options, we accumulated an enormous debt of gratitude to those who gave so freely of their time and knowledge to assist us along the way. Their scientific curiosity and passionate interest in the culinary arts have inspired and guided us in the process of composing and writing this book.

Of the many individuals who put technical and professional knowledge at our disposal, cheerfully participated in our experiments, and facilitated our expeditions around the world to seek out umami, particular thanks are due to: the fascinating people who gather together as The Funen Society of Serious Fisheaters and The Dozen Society, who helped to shape our sensitivity to umami from the pantry in the sea; our good friend and fish expert Poul Rasmussen, for enjoyable and inspiring conversations and gastronomical experiments with fish, shellfish, *ikijime*, clambakes, and fish sauce production; and the chefs Torsten Vildgaard, Lars Williams, and Søren Westh from Restaurant noma and Nordic Food Lab, and the chefs Pepijn Schmeik and Remco van Erp from Restaurant Eendracht for providing insight into their playful, yet serious, approach to culinary adventures.

Thanks also are due to: Yukari Sakamoto, for carefully scrutinizing the Japanese expressions; Dr. Carl Th. Pedersen, for advice with respect to the chemical and gastronomic expressions in the book; Dr. Niels O. G. Jørgensen and Lars Duelund, for measurements of glutamate in a large number of samples; wine experts Peter Winding and Pia Styrbæk, for tastings and enlightening discussions regarding wine pairings for dishes with umami; Dr. Ling Miao, for information on Chinese soups and help with Chinese quotes; Professor Ylva Ardö, for information on maturation of cheeses; Ayako Watanabe, for pointing out references to data for the amino acid content of sake and for conversations together with chef Yoshitaka Onozaki about *shōjin ryōri*; chef Hiroaki Yamamoto for information on *kobujime*; Dr. Christian Aalkjær, for information about salt and blood pressure; chef Søren Gordon from bar'sushi, for preparing *gunkan-zushi* for photography; Sakiko Nishihara, for information about Taste No. 5; Pierre Ibaïalade Co., for a guided tour of its facilities for salting and drying hams in Bayonne, France; Dr. Lee Miller, for supplying *kusaya*; Reidun Røed and Martin Bennetzen, for providing Norwegian *rakfisk*; Dr. Jorge Bernadino de la Serna, for samples of Spanish *botargo*; brewer Ole Olsen, for information about free amino acids in beer; Henrik Jespersen, for information about *rakfisk*; Dr. Søren Mørch, for participating in experiments on *ikijime* and the preparation of a clambake; and Dr. Michael Bom Frøst, for valuable background about sensory sciences.

We would also like to thank Dr. Kumiko Ninomiya, for useful information on dashi preparations, Japanese fish sauces, and umami compounds in soup broths, and for making available the original writings of Kikunae Ikeda, as well as unpublished data on glutamate content in *ichiban* dashi. For their hospitality, we would like to thank the following people: Dr. Koji Kinoshita, for help and guidance during a visit to the Osaka area, for valuable information about Japanese traditions and food culture, and for advice regarding the Japanese version of the quotes by Kikunae Ikeda; Drs. Kumiko Ninomiya, Ana San Gabriel, and Kazuya Onomichi, as well as other members of the Umami Information Center and the Ajinomoto Research Laboratories, for outstanding hospitality when Ole visited Tokyo in 2013 and for arranging a tour to inspect *katsuobushi* production in Yaizu; Tooru Tomimatsu, president of Katsuo Gijutsu Kenkyujo, for a guided tour of the harbor and *katsuobushi* production facilities at the company Yanagiya Honten in Yaizu; Saori Sawano, owner of the wonderful knife store Korin in New York, for kindly mediating a contact with the Sakai City Industrial Promotion Center in Osaka, Japan; Tsutomu Matsumoto, who showed Ole around at the seaweed production company Matsumoto in Sakai and provided valuable information on konbu quality and storage conditions for optimizing umami; and Hiroki Yamanaka, who guided Ole on a tour of seaweed production sites in Sakai. Finally, for help with recipes, we thank the following people: Kirsten Drotner, for the recipe for green pea soup; Inger Marie Mouritsen, for the recipe for traditional spiced pork sausage; Kristin Lomholt, for the recipe for a dressing with nutritional yeast; Larissa Zhou, for imaginative contributions to the Nordic dashi project; and chef Yoshitaka Onozaki, for the recipe for 'fried eel' made from lotus root (*kabayaki*).

We wish to express our sincere gratitude to chef Israel Karasik from Restaurant Kvægtorvet in Odense, Denmark, for being a unique source of analytical and technical inspiration during the development and testing of the new recipes presented in the book. Moreover, we wish to extend our thanks to the other chefs at Kvægtorvet, for their patience and invaluable help during tastings and experiments.

Ole wishes to acknowledge a special grant from VILLUM FONDEN, which enabled him to carry out pilot projects regarding seaweeds and taste. He also benefited greatly from the Palsgaard Estate's generous loan of Stinnes Hus in As, which provided him with a tranquil escape for a period of intense writing.

Much of the factual information on which the book is based is found in the references listed in the bibliography. Moreover, the Umami Information Center and the book *Dashi and Umami: The Heart of Japanese Cuisine* have been important sources of inspiration and data.

This book was originally written and published in Danish, the mother tongue of the authors. The present English edition is a fully updated and revised version of the Danish work, translated and adapted into English by Mariela Johansen. Mariela enthusiastically undertook the ambitious task of turning the interdisciplinary material into the coherent, scientifically sound, and very readable book you now hold in your hands. She did an admirable job not only of translating the book but also of checking facts, ensuring consistency, and suggesting new material and valuable revisions. The authors owe much to Mariela for caring so much for the project.

The format, layout, and graphics were all designed and executed by Jonas Drotner Mouritsen. Jonas has been a crucial participant in the project from the beginning. It is due to his creative skills that the text, photographs, and other illustrations were integrated so successfully. Figures and photographs made available by a number of individuals and organizations greatly enhance the book. A list of these contributors can be found at the back of the book.

Finally, we are indebted to our editor, Jennifer Crewe, for her enthusiastic support of the project, and Columbia University Press for professional and expeditious handling of the manuscript.

Ole G. Mouritsen and Klavs Styrbæk
Odense, Denmark

Prologue: How it all began

Some readers might be curious to know a little about what inspired us to undertake this joint venture to unlock the secrets of umami and to put our findings together in a book. Like most Danes, we were very familiar with the four basic tastes, enshrined in Western literature for many centuries: sour, sweet, salty, and bitter. But the idea of a 'fifth taste,' one that has been known in the East for millennia, had not gained much traction in the circles we frequented, even though the popularity of Asian food had grown by leaps and bounds in the past few decades. In fact, the concept of the fifth taste, umami, which roughly translates from the Japanese as 'deliciousness,' had not really started to be associated with other cuisines.

In a nutshell, this fairly closely describes our own relationship to umami and how this led to an unusual collaboration. A few years ago, we, Klavs and Ole, had both been invited to speak at an evening event that was part of a series of informal, university-style lectures for the general public. We had both just published books—Klavs had written about what he calls 'grandmother's food,' old-fashioned Danish cuisine, and Ole had just finished a broadly based book on seaweeds, including its underexploited potential as food. As part of his talk, Klavs had prepared a tasting menu in which he had replaced the bacon in a very traditional dish with a seaweed, dulse. In the course of the presentation, he uttered the word *umami*, not exactly an expression that was common in our native Denmark and certainly not one that was associated with Danish food. Ole already knew about this mysterious fifth taste from a decades-long love affair with Japanese cuisine and, more recently, his interest in it as it related to edible seaweeds. When Ole approached Klavs afterward to ask what the term 'umami' meant in his universe, that of gastronomy, the idea of writing a book together was floated, and the project soon took on a life of its own.

There is something truly exciting about running up against a challenge to our preconceived notions of the world and how it is organized. These ideas have often developed gradually and imperceptibly in the course of our lives without our even being aware of their presence. But if we are suddenly confronted with a reality that does not align with our outlook, or that perhaps is much bigger and more all-encompassing than we had believed, it can lead to one of those famous 'aha!' moments. We start to become aware of details we had not noticed before or, possibly, knew about but had not really articulated as a concept with a distinct name. We discovered that umami was as deeply embedded in European cuisines as in those of the East. By attaching a single word to this taste, we were immediately able to bring into focus a host of discrete sensory impressions related to it and to start to analyze them.

We approached the subject from very different perspectives. As a professional chef, Klavs sees great value in the venerable traditions of Danish food culture, while at the same time exploring ways in which it can be renewed by taking advantage of modern food science and the precepts of the New Nordic Cuisine, which emphasizes local, seasonal products of the highest quality. Ole, on the other hand, is a research scientist focusing on the discipline of biophysics, who is also an amateur chef with a great deal of curiosity about food at the molecular level and who enjoys sharing his knowledge as widely as possible. In a sense, our collaboration has had parallels with how umami works. As you will soon learn, the taste can be imparted by two different types of substances, glutamate and nucleotides, which can interact synergistically to enrich its effect beyond the contributions made by each type of substance. In relation to this book, our two distinct but complementary skill sets helped us to achieve more together than we could have simply by compiling our individual efforts.

A WORD ABOUT RECIPE MEASUREMENTS

Quantities for ingredients are given in both metric and imperial units, bearing in mind that conversion from the one to the other can only be approximate. Usually this is not an issue, as few of us prepare meals by weighing out ingredients to the nearest fraction of a gram or by using laboratory equipment to measure a liquid. We generally know what is meant by a cup and a teaspoonful, and greater accuracy is normally not needed. Many of the recipes in the book are of this type. In a few instances, where very precise, small quantities are indicated—for example, for yeast—it is important to pay close attention.

▸ The chef in the kitchen.

This volume is not intended to be only a cookbook, but is also meant to be a source of information that will foster a greater awareness of umami and allow readers to kick-start their own ideas about how they can take advantage of the benefits it offers. To that end, we have included a number of simple recipes and practical tips along the way. We have also included a small selection of recipes that are of a whole different level of complexity and that are intended to be inspirational and aspirational. While readers may not have the equipment or patience to try these recipes, we feel that they have a role to play by generating 'aha!' moments that will translate into adapting ideas from these dishes for use in everyday meals.

It is our hope that this book may serve as an eye-opener for a diverse audience—those who write about food, professional cooks, and engaged readers—and lead them to marvel at the mysteries inherent in the culinary arts and to ask a few questions about what might lie behind the small miracles of taste. Armed with some basic knowledge about how umami works and where to find it in raw ingredients, all readers should be able to use the information to unleash their creativity and invent their personal, signature umami dishes—in other words, to unlock the secrets of the fifth taste.

*When there is
no longer good cooking in
the world, there will be no
literature, no great intelligence
and quick wit, no inspiration,
no friendly gatherings,
no social harmony.*

*Lorsqu'il n'y a plus de cuisine dans le monde,
il n'y a plus de lettres, d'intelligence élevée et
rapide, d'inspiration, de relations liantes,
il n'y a plus d'unité sociale.*

Marie-Antoine Carême (1783–1833)

What exactly is taste, and why is it important?

Well into the modern age, taste was regarded as something subjective over which housewives and chefs held sway. It was not until about the 1920s that it became the object of rigorous scientific studies. So it should come as no surprise that it is only within the past few decades that we have started to gain a better understanding of its actual physiological basis. This allows us to explain, in detail, how taste is detected in the mouth by certain receptors and converted into nerve impulses that are then forwarded to specific centers in the brain. The neural cells in these centers carry out the final calculation and convey a message about the food—for example, sweet! or salty!

THE BASIC TASTES: FROM SEVEN TO FOUR TO FIVE AND POSSIBLY MANY MORE
For many centuries, it was customary in the Western world to accept the ancient Greek view, originating with Aristotle, that there were seven basic tastes: sour, sweet, salty, bitter, astringent (causing dryness), pungent (or spicy), and harsh. Over time, people came to the conclusion that there were actually no more than four true basic tastes; namely, the first four on this list. But it was only in the course of the twentieth century that a clear distinction was drawn between sour, sweet, salty, and bitter as genuine tastes and the other three as mechanical or chemical effects caused by substances in the food that damage the cells on the tongue or in the mouth.

In many countries in Asia, however, people have thought that in addition to these four basic tastes, there is a fifth one—pungent or spicy, for example, as in chile peppers. Complicating matters, according to classical Indian philosophy, astringent is also a separate taste. On the other hand, in China and Japan, there is a long-standing tradition, possibly going back more than a thousand years, that there is a particular, identifiable taste associated with food that is especially delicious. In 1909 this taste was given the name *umami*, a new Japanese word combining the ideas of *umai*, which means 'delicious,' and *mi*, which means 'essence,' 'essential nature,' 'taste,' or 'flavor.' While some Japanese are not overly familiar with this term per se, many others use it not only to denote a mere taste but also as an expression for that which is perfect.

There is no single word in Western languages for this particular taste, nor for a sensation of taste, that is equivalent to how a Japanese person experiences umami. Perhaps this is because the concept of umami is not associated with a universally known and well-defined source in Western cuisines, unlike, for instance, the identification of table salt with saltiness, sugar with sweetness, quinine with bitterness, and vinegar with sourness. In the Japanese kitchen, there is a single ingredient, with a very pure taste, that quintessentially typifies umami—this is the traditional and ubiquitous soup stock dashi, which is used not only in soups but in many other dishes. While there is a great deal of food in the West that is characterized by umami, it is often found in combination with other tastes, for example, in complex mixtures of meat and vegetables, which may also contain considerable quantities of oils and fats. The result is a pleasant, but also more complicated, taste impression. Consequently, if they think about it at all, Westerners tend to view umami as merely a new word for an old, familiar set of taste sensations.

It seems, however, that the Chinese and Japanese have been right all along, as it has now been scientifically established that there are actually five different basic tastes. Of these, the umami, sweet, and bitter ones are the most important in determining how we react to particular foods. Foodstuffs with a sweet or umami taste are generally considered agreeable, while those that are bitter are often rejected.

All of this brings us back to some fundamental questions: What exactly is taste, how do we experience it, and why is it important?

A *taste* is a sensory impression to which, in principle, we can assign an objective biochemical and physiological perception of a substance; let us just say a molecule, whose chemical nature determines its taste for us as humans. It is not a given that another animal—for example, a mouse—would discern it in the same way.

The *experience of taste* is much more involved than the physical perception of taste and is often quite particular to an individual. Although it is a function of the same biochemical processes as taste, it is also influenced by the other senses: sight, sound, the feel of the food in the mouth, and especially our sense of smell, which is much more discriminating than that of taste. In addition, the gustatory experience is affected by psychosomatic conditions, social context, cultural background, traditions, degree of familiarity with the food, and, finally, whether we are hungry or already feel full.

There are many types of taste, and a human may possibly be able to distinguish between several thousand different ones. An overall taste is typically made up of a small number of basic tastes. From a scientific perspective, in order for a taste to be considered a true basic one, it must be independent of all other basic tastes and, at the same time, be universally present in a wide variety of foods. In addition, a basic taste must be the result of a physiological phenomenon that, in turn, depends on a chemical recognition of the taste. This recognition takes place with the help of particular proteins, known as taste receptors, which are found in the taste buds on the tongue. It has been known for many years that there are special receptors for the sour, sweet, salty, and bitter tastes. The first receptor for one of the substances that imparts umami, namely, an amino acid (glutamic acid) and its salts (glutamates), was discovered in 2000. As a result, umami could justifiably be elevated to the status of a true basic taste, 'the fifth taste.' Subsequent studies have identified additional receptors for umami.

What is interesting about pure glutamate in the form of monosodium glutamate (MSG, sometimes called the third spice), is that it cannot really be said to be tasty on its own. Rather, one might say that MSG has no taste or, even worse, that it tastes like a mixture of something salty, bitter, and maybe soapy. It is only in combination with other taste substances that it calls forth that sublime taste sensation that is worthy of the splendid name *umami*. For this reason, MSG is often characterized as a taste enhancer. It interacts strongly with other common taste substances, especially table salt, NaCl. What is distinctive about MSG is the nonlinear synergy between it and other substances that also impart umami—a very small quantity of these other substances, known as 5'-ribonucleotides, has a notable multiplier effect on the action of the MSG. As a result, there are many as-yet-unimagined possibilities for playing with umami by combining a range of different raw ingredients.

So even though unique words for umami are lacking in the vocabularies of Western languages, this taste has, of course, not been absent in our kitchens. When examined more closely, traditional European cuisines are seen to strive as much to incorporate umami into their dishes as do the Asian cuisines. Soups based on meat and vegetables, cured hard cheeses, air-dried hams, fermented fish, oysters, and ripe tomatoes are all evidence that we crave after, and savor, foods that are rich in umami tastes.

The science underlying food is complex. Our sensory apparatus for tasting and enjoying food is equally complex and, in many ways, poorly understood. In fact, the sense of taste is the least well understood of the human senses. It is not a given that all taste impressions can be described using only five elementary types of basic tastes. It is conceivable that there might be more than five. Some researchers have recently published studies indicating that they have found a fat receptor in the taste buds on the tongue, suggesting that fattiness might be a basic taste.

WHY DO WE NEED TO BE ABLE TO TASTE OUR FOOD?
In a modern society where there is an abundance of food, we probably think of taste as something that primarily adds sensual pleasure and delight to the enjoyment of a meal. Some might even think of an appetizing taste as something that induces people to be bothered to eat at all. The majority of us, who are not engaged in hard physical work, are not really hungry when we eat. To be convinced of this, just reflect on how much a hiker looks forward to digging into a simple bag lunch during a rest stop in the middle of a strenuous mountain trek.

It is likely that taste allows an animal species to identify those foods that help to ensure its survival, as well as those that might be harmful. This could confer certain evolutionary advantages, although it is, admittedly, difficult to prove this hypothesis. It is evident, however, that the evolutionary basis for taste is probably not sensual pleasure, but rather a fulfillment of a fundamental need and the will to survive and reproduce. To this end, the individual needs food that is very nutritious (proteins), food that provides energy (calories from fats and carbohydrates), and food that contains salts and minerals. In addition, taste has to indicate whether or not the food is poisonous. In all likelihood, the basic tastes have, since time immemorial, been signals that show us how to meet these fundamental nutritional requirements.

What do the various basic tastes tell us?

- *Sweetness* tells us that the food contains sugars and the metabolic by-products of carbohydrate breakdown, which provide energy and calories.
- *Saltiness* indicates the presence of minerals and salts, such as those from sodium and potassium that are vital for preserving a proper electrolyte balance in our cells and organs to ensure their proper functions.
- *Bitterness* sends a strong message that the food may contain poisonous substances—for example, alkaloids—that we should avoid.
- It is less obvious why we taste *sourness*. Acidity might steer us toward substances that regulate the pH balance in our bodies while at the same time sharpening the appetite and improving digestion. At any rate, sourness helps us to stay away from foods, such as unripe fruit or rancid fats, that contain so much acid that they can be unpleasant to eat or even poisonous.
- If it should prove to be correct that there are also specific receptors for *fattiness*, it would presumably be a sign that the food contains a significant energy supply.
- In all likelihood, we can taste *savoriness* or umami because it tells us that the food contains readily accessible nutrition in the form of amino acids and proteins. And furthermore, the intensity of the umami taste gives us an indication of how ripe and full of nutrition a particular food might be. It is quite possible that we are genetically programmed to enjoy umami.

Along the same lines, one might be able to say that the drive to find food that tastes good and that is rich in umami makes *Homo sapiens* a gourmet ape.

THERE IS MORE TO IT: SENSORY SCIENCE, TASTE, SMELL, AROMA, FLAVOR, MOUTHFEEL, TEXTURE, AND CHEMESTHESIS

The study of our perception of food, especially of taste, is known as sensory science. Rather than the word *taste*, we should instead use the word *flavor*, which denotes the integrated effect of all sensory impressions evoked in the oral cavity. It encompasses both taste and smell, including those derived from aromatic substances in the food, as well as mouthfeel and chemesthesis, which is a sense category that relies on the same receptor mechanisms as those that convey pain, touch, and temperature in the eyes, nose, mouth, and throat.

TASTE OR FLAVOR?
In ordinary speech, the terms taste and flavor are often used interchangeably, but strictly speaking, they are quite different. A *taste* has to fall into one of the known classifications for which there are distinct taste bud receptors. *Flavor*, on the other hand, is a perception based on three essential elements: the combination of tastes in the food (think, for example, of real black licorice, which is both salty and sweet), the effect of the aromatic components on the olfactory receptors, and the feelings related to texture, temperature, and so on that are evoked in the mouth.

ODOR, SMELL, OR AROMA?
The words *odor*, *smell*, and *aroma* can all be used to denote that which we perceive through the olfactory system. Although the words in themselves are neutral, odor and smell tend to have a negative connotation. An aroma is also a smell, but the word is used to signify that it is a pleasant one, usually associated with food.

A RECENT ARRIVAL ON THE SENSORY SCENE: *KOKUMI*
The Japanese expression *kokumi* (derived from *koku*, meaning 'rich' and *mi*, meaning 'taste') was coined a few years ago by researchers at the Japanese company Ajinomoto. It combines three distinct elements: thickness—a rich, complex interaction among the five basic tastes; continuity—the way in which long-lasting sensory effects grow over time or an increase in aftertaste; and mouthfeel—the reinforcement of a harmonious sensation throughout the whole mouth. It has been shown recently that *kokumi* is evoked by the stimulus of certain calcium-sensitive channels on the tongue by small tripeptides (for example, glutathione) found in foods such as scallops, fish sauce, garlic, onions, and yeast extract. Whereas glutamate has a significant effect on the umami taste in concentrations of about one part per thousand, substances that produce the most potent *kokumi* need to be present in concentrations of only two to twenty parts per million.

Because an individual's experience of flavor results from a very complex combination of several types of sensory perception, it is not always easy to relate a given flavor to the chemical composition of the food.

The sensation of taste presupposes that the taste substances are dissolved in a liquid, primarily in the mouth. As already mentioned, its perception is mediated by the taste receptors, which are located in the taste buds on the tongue.

The sense of smell depends on airborne substances in the form of single molecules, particles, or vapor droplets. These are either released in the oral cavity when the food is chewed and work their way internally to the nasopharynx (retronasal stimulation) or are given off by the food and inhaled through the nostrils (orthonasal stimulation). Smell by the retronasal route appears to be the more important for humans, whereas the opposite is true for dogs. In both cases, the aromatic substances reach the roof of the nasal cavity, where there is an array of specialized neural cells located under a mucus membrane that is covered with tiny cilia. Here sensory cells with olfactory receptors, of which there are thousands of different types, can detect them. As any particular odor generally activates several receptors, humans are able to distinguish among a vast number of different smells. The sense of smell is much more fine-tuned than that of taste, and is now believed to form a sensory image in the brain.

Mouthfeel is a collective term for the sensory perceptions that are neither taste nor aroma but that interact closely with them. It is influenced by the structure, texture, and morphological complexity of a food item and is, to a great extent, responsible for our overall impression of the food. For example, this can involve physical and mechanical impressions such as chewiness, viscosity, mouthcoating, and crunchiness.

The Japanese have a special expression, *kokumi*, which is rather difficult to convey in other languages. It encompasses thickness, continuity, and mouthfeel, and may overlap somewhat with the taste sensations evoked by umami. *Kokumi* is not an independent taste, but it does refer to taste enhancement and is associated with food that is truly delicious.

Chemesthesis is a technical term that describes the sensitivity of the skin and mucus membranes to chemical stimuli that cause irritation. It can be thought of as an early warning that these may be harmful. An example of chemesthesis is the painful burning sensation on the tongue that we associate with sharp or spicy tastes caused by a variety of substances such as those in chile peppers (capsaicin), black pepper (piperine), and mustard (isothiocyanate).

Thermal perception of warmth and temperature in the mouth is related to chemesthesis. It is based on the chemical activation of six different temperature-sensitive ion channels located in the membranes of the sensory cells. This sense is so finely tuned that we are able to detect temperature fluctuations to within 1 degree. If the temperature of a substance is less than 15°C or more than 43°C, we experience it as pain. Some chemical substances can fool this sensory system and activate the ion channels directly, leading us to think that a taste experience is warm or cold, even though the temperature is actually unchanged. This is referred to as a false perception of heat or cold. For example, we experience capsaicin from chile peppers as hot and menthol, peppermint, and camphor as cool.

A more mechanical sensory impression is that of astringency, which we know from the taste, for example, of tea or red wine, both of which are rich in tannins. It is caused when certain chemical substances interact with proteins found in the mucus on the surface of the tongue and in saliva. It is described as causing feelings of sharpness, dryness, and friction.

IS THERE A TASTE MAP OF THE TONGUE?
Since the early 1900s, it has been commonly believed that the threshold for detection of the different basic tastes varies across the tongue and that the experience of each of the tastes is exclusively localized to a distinct area on it. This concept, which turns out to have been mistaken, is derived from subjective impressions that we taste sweetness at the tip of the tongue, saltiness at the sides toward the front, sourness also at the sides but further back, and bitterness at the root of the tongue. Seemingly, there is an area in the middle of the tongue where we feel that there is a decreased sense of taste.

More recent scientific research has shown that this so-called taste map is incorrect. The different regions of the tongue are sensitive to all the basic tastes, although they may perceive them to varying degrees.

Controlled experiments to determine precisely which areas of the tongue are most sensitive to umami have identified the part around its root as the area of greatest sensitivity. Nevertheless, when research subjects are asked where they taste umami, they generally answer that they taste it

Sour

Sweet

Salty

Bitter

Umami

The taste map. Schematic illustration of the areas on the tongue, indicating the location of the greatest number of taste buds and taste receptors. The five basic tastes are all detected in each of the areas.

everywhere on the tongue. This indicates that the subjective taste sensation of umami is not always in accord with the physical distribution of the specific receptors for different tastes. In all likelihood, this explains why umami is often perceived as a wall-to-wall taste experience that completely fills the mouth with delicious sensations.

This particular way in which we experience umami may be one of the reasons why people in the West have been so slow to accept it as a true basic taste. Some chefs think that Westerners have a sort of serial experience of taste, in which the different taste sensations and nuances are perceived by way of contrasts and complementarities in a linear, stream-like fashion, whereas Asians take them in all at once and process them in parallel. As a result, umami can possibly be regarded as a parallel or complete taste.

WHY ARE SOME FOODS MORE PALATABLE THAN OTHERS?
Taste and, to a much greater extent, the sensation of taste, both of which relate directly to palatability, have a subjective and psychological component that puts all of our senses into play. Palatability is central to our choice of food, as well as to how it is processed and digested in our body. Our experience of palatability typically is a combination of many factors. The brain carries out the final assessment of these and tells us whether or not a particular food tastes good.

How these many complex impressions combine and affect one another is of special importance for our understanding of the relationship between palatability and umami. Knowing something about these interactions will help us to understand the nature of umami, and, in addition, enable us to work out distinct ways of enhancing this taste in our own cooking.

A particular aspect of what makes umami delicious is aftertaste. Umami develops over a different time frame than do saltiness and sourness, which disappear quite quickly. Experiments have shown that the intensity of those substances in the food that bring out umami actually increases for a short period of time after the research subject has spit out or swallowed the food. Umami persists for longer than all the other basic tastes.

This lingering aftertaste is probably one of the reasons why we associate umami with deliciousness and something pleasant. It is a taste sensation with fullness and roundness that completely permeates the oral cavity and then dissipates very slowly.

Air-dried hams are rich in glutamate, which brings out an abundance of umami.

It is probably easiest to describe what umami is by talking about its absence, which leaves us with food that we characterize as boring, flat, and uninteresting. In this book, we will describe how the richest and most delicious umami tastes arise when certain substances are present in particular combinations. We have only to take the trouble to develop expertise in combining different ingredients and handling them in the right way to be able to tease out their inherent taste substances.

A FEW WORDS ABOUT PROTEINS, AMINO ACIDS, NUCLEOTIDES, NUCLEIC ACIDS, AND ENZYMES

Proteins and amino acids have a special role to play in this book because they are the main sources of umami. Proteins are composed of amino acids, which are small molecules that can bind chemically to each other with what are known as peptide bonds to form long chains of molecules. Some protein molecules are extremely long, made up of as many as a thousand amino acids. An example of this type of protein is wheat gluten, which, as its name implies, contains a great deal of the amino acid glutamic acid.

Our food contains twenty different naturally occurring amino acids. Nine of them are called essential amino acids because our bodies cannot synthesize them and, therefore, must derive them from what we ingest. Glutamic acid, the source of umami, is not one of these, and our bodies can produce it, even in great quantities.

Amino acids are chiral molecules, meaning that they are found in two versions that are chemically identical but are mirror images of each other, like the right and left hands. One of the properties that distinguishes them from each other is how they rotate a plane of polarized light that is passed through them. Those that rotate it counterclockwise are called levorotatory or left turning (L-amino acids), and those that rotate it clockwise are known as dextrorotatory or right turning (D-amino acids). The direction in which the amino acid turns can lead to differences in taste.

Amino acids can form salts with, for example, sodium, potassium, magnesium, calcium, or ammonium. We have already come across the sodium salt of glutamic acid, which is monosodium glutamate (MSG).

Proteins are important in a nutritional context, because they provide some of the building blocks and energy necessary for cellular function. On their own, large protein molecules are rather insipid, whereas they can make a major contribution to how food tastes when they are broken down into small peptides or free amino acids. Knowing how best to break proteins down to free amino acids, usually by cooking, fermenting, curing, drying, marinating, or smoking, is an essential aspect of the culinary arts. Many free amino acids taste bitter, and many are predominantly sweet. (See the tables at the back of the book.) Some that are sweet actually taste bitter in large quantities.

Two amino acids are sources of umami taste: primarily glutamic acid in the form of glutamate and, to a considerably lesser extent, aspartic acid in the form of aspartate. For example, monosodium aspartate (MSA) imparts umami, but the effect is only 8 percent of what can be achieved with glutamate. Only a small portion of the glutamic acid in the protein content of fresh food is found in the form of free amino acids. Furthermore, only the free glutamate ions and aspartate ions, rather than the amino acids themselves, result in umami.

> ### GLUTAMIC ACID, GLUTAMATE, AND THE GLUTAMATE ION
>
> In this book, we will use these three terms in connection with descriptions of umami. Which one we use will depend on the context. Normally, we will discuss the amino acid glutamic acid in connection with its presence in proteins, where it is bound to many other amino acids. In this bound form glutamic acid has no taste. By appropriate processes, it can be liberated from the proteins and act as a free amino acid when dissolved in water. As long as the glutamic acid is in the form of an acid (for example, in a sour solution), it does not give rise to umami. On the other hand, if it forms a salt by combining with another compound (for example, sodium), the glutamic acid takes on the form of glutamate, in this case monosodium glutamate (MSG). In solution, this salt separates into sodium ions and glutamate ions. The glutamate ion stimulates the glutamate receptor and produces the umami taste.
>
> So it is not the actual MSG that results in the umami taste, but only the glutamate ion. For the sake of convenience, we will also refer to the glutamate ion as glutamate, depending on the context in which the glutamate imparts umami. The word *glutamate* will therefore be used to describe free glutamic acid that has formed a glutamate ion. In this sense, all the glutamate that will be discussed is really free glutamate, which can be perceived by the glutamate receptor.

Nucleotides are molecular groups that can bind together in long chains (polynucleotides) and form nucleic acids, such as RNA or DNA, which are the foundations of our genome. With regard to umami, it is particularly the 5'-ribonucleotides derived from inosinic acid, guanylic acid, and adenylic acid—namely, inosinate (IMP), guanylate (GMP), and adenylate (AMP)—that are important, as they interact synergistically with glutamate to increase umami.

ATP (adenosine-5'-triphosphate), which is the primary biochemical energy source in living cells, is another important polynucleotide. When it is broken down it can form, among other substances, the three 5'-ribonucleotides mentioned above that are linked to the umami taste.

In contrast to proteins, nucleic acids are not in and of themselves nutritionally important, but the free nucleotides formed as by-products of their breakdown can act to increase umami. Furthermore, recent research has, surprisingly, shown that even though our body can synthesize the nucleotides that it needs, the free nucleotide content found in our food intake seemingly plays an important role in building up the immune system, especially in the intestines of newborn babies. This possibly explains why human breast milk contains so many free nucleotides.

A particular type of proteins called enzymes can break other proteins or nucleic acids down into their constituent parts; that is to say, either into free amino acids or free nucleotides. This is where taste comes into the picture, because we can taste them in this form, even though we cannot taste either the proteins or the nucleic acids from which they are derived.

GLUTAMIC ACID AND GLUTAMATE IN OUR FOOD
Because glutamic acid is such a vital building block in proteins, it is found in large quantities in many of our foodstuffs, in either bound or free form. It makes up 10–20 percent by weight of animal proteins and as much as 40 percent by weight of plant proteins.

In the animal kingdom, glutamic acid is found in meat, poultry, and fish, while in the plant kingdom, it is abundant in vegetables but occurs only in small quantities in fruits. Vegetables are characterized by a relatively large content of free glutamate; for example, in tomatoes, corn, potatoes, and peas. (See the tables at the back of the book.) From the third major kingdom, the algae, which are not yet eaten very widely in Western countries, we obtain large quantities of free glutamate from, among others, large brown marine algae (seaweeds) such as konbu (*Saccharina japonica*), which is used to make the Japanese soup stock dashi.

On average, persons living in the Western world ingest about 30 milligrams per kilogram of body weight of free glutamate from their regular daily food intake. This corresponds to about 2 grams daily for an adult. One can also factor in an additional 0.3–1 gram daily sourced from additives. In many countries in the East, such as Korea and Japan, the daily intake of glutamate from additives is up to three times as great.

It is important to be aware that a given foodstuff can have a relatively low free glutamate content compared to another food, but at the same time have a relatively high content of bound glutamic acid, or vice versa. For example, cow's milk has very little free glutamate but a quite large amount of bound glutamic acid. Consequently, fresh cow's milk does not have much umami, whereas fermented milk products, such as aged cheeses, are good sources because glutamate was released in the course of the fermentation process.

HOW DOES GLUTAMATE TASTE, AND HOW LITTLE IS REQUIRED FOR US TO TASTE IT?

It is very difficult to carry out objective, quantitative measurements of taste perceptions, and the results of experiments depend to a great extent on the methods that are employed to do so. In this connection, it is important to understand that both the taste threshold and taste intensity come into play. *Taste threshold* is an expression for the minimum quantity of a substance that is needed in order for us to perceive its taste. Determining a parameter for *taste intensity* is more problematic, as it is highly subjective.

Experiments have shown that the lower limit for tasting MSG in pure water is 0.01–0.03 percent by weight. (See the tables at the back of the book.) As mentioned, however, this threshold is very dependent on the method used to measure it. The equivalent threshold for table salt in pure water is about twice as high. As we will see later, the taste threshold for umami can be hundreds of times lower if other substances, such as inosinate, that enhance this taste are also present. The taste intensity of glutamate increases logarithmically with the concentration, but it has a tendency to saturate. It should be noted, however, that umami in foodstuffs is normally a mild and subtle taste, not nearly as intense as that which we associate with sweet and sour ones found in honey and lemons, respectively.

In a typical soup, there needs to be about 10 grams of salt per liter for it to taste sufficiently salty, and a reasonably narrow range of 8–12 grams per liter determines whether the soup comes across as insipid or too salty. In the case of MSG, a relatively broader range of 1–5 grams per liter ensures that it tastes good. The optimal salt content in a dish will decrease when MSG is also present, just as nucleotides depress the threshold for the optimal MSG concentration. (See the tables at the back of the book.)

What is probably surprising is that the taste of pure MSG is neither particularly pleasant nor interesting. In fact, it is rather bland and somewhat soapy. Its taste is perceived as delicious only when it is eaten in combination with a variety of foods. Here we are getting closer to what umami is all about—it is not the taste of pure glutamate. It is a much broader concept.

*In the same way
as the colors arise from the
mixture of white and black, the tastes
arise from mixtures of the sweet and the
bitter.... Richness is an aspect of the sweet
taste...while the salty and the bitter are nearly
the same. Between the extremes of the sweet
and the bitter come the harsh,
pungent, astringent, and
acidic tastes.*

ὥσπερ δὲ τὰ χρώματα ἐκ λευκοῦ καὶ μέλανος
μίξεώς ἐστιν, οὕτως οἱ χυμοὶ ἐκ γλυκέος καὶ
πικροῦ, καὶ κατὰ λόγον δ' ἢ τῷ μᾶλλον καὶ
ἧττον ἕκαστοί εἰσιν, εἴτε...ὁ μὲν οὖν λιπαρὸς
τοῦ γλυκέος ἐστὶ χυμός, τὸ δ' ἁλμυρὸν καὶ
πικρὸν σχεδὸν τὸ αὐτό, ὁ δὲ δριμὺς καὶ
αὐστηρὸς καὶ στρυφνὸς καὶ ὀξὺς ἀνὰ μέσον

Aristotle (384–322 BCE)

The first four: Sour, sweet, salty, and bitter

The perception of taste has its physiological origin in the taste receptor cells. These are found in the taste buds, which are embedded in tiny protrusions (papillae) located primarily on the top of the tongue but also distributed over the soft palate, pharynx, epiglottis, and the entrance to the esophagus. There are approximately 9,000 taste buds on the human tongue, clustered together in groups of 50 or so. Each taste bud is made up of 50–150 taste receptor cells. Like other cells, they are encapsulated in a cell membrane, and it is this membrane that holds the secret of taste perception.

THE PHYSIOLOGY AND BIOCHEMISTRY OF TASTE
The taste receptor cells are bundled together in onion-shaped taste buds in an arrangement that fits together much like the segments of an orange, forming a small pore at the surface of each papilla. Small hair-like structures (microvilli) extend from the cell membranes at the top of these taste receptor cells. Taste substances must pass through the pores in order to be detected, and they can do so only if they are dissolved in water or saliva.

The receptors that can identify the taste substances biochemically are located in the membranes of the taste cells. A receptor is a large protein that functions as a sort of antenna that normally can detect and identify only one type of taste. But in order to be detected by the receptor, the taste molecules generally have to be small. The majority of pure carbohydrates and proteins have absolutely no taste because their molecules are just too large.

In the past few years, major advances in the field of molecular biology have provided us with a much more detailed insight into how these receptors look at the molecular level and how they function.

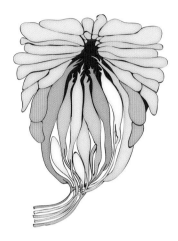

Schematic illustration of a taste bud. The taste bud is covered by some epithelial cells that form a little pore, through which the taste substance can enter the taste bud. The taste bud consists of a bundle of taste receptor cells, which are individually responsive to one basic taste: sour, sweet, salty, bitter, or umami. Cells that detect the same tastes transmit the information through a single nerve fiber that forwards the collective signal to the brain, which then registers the taste sensation.

A number of models have been proposed to explain how the various taste receptor cells transmit their signals from the taste buds. There now appears to be a consensus around a model in which each taste receptor cell is primarily adapted to detect one particular taste. The various cells that are sensitive to the same taste collect the signals in a single nerve fiber that sends the total signal onward through the cranial nerves to the taste center in the brain. The picture is complicated by the presence in the taste buds of another type of cells, known as the presynaptic cells, which have no taste receptors but still participate in the transmission of the nerve signals. In contrast to the taste receptor cells, the presynaptic cells can respond to several different types of taste, given that they receive signals from a number of taste receptor cells.

There are two principal types of taste receptors. One type, called the G-protein-coupled receptors, which are sensitive to sweet, bitter, and umami tastes, traverse the membrane of the sensory cell. When such a receptor on the surface of the cell has identified a taste molecule for which it is adapted and has bound it, a signal is passed through the protein that a certain other protein (known as the G-protein), located on the other side of the membrane, is also to be bound. This binding sets in motion a cascade of biochemical processes that eventually cause particular sodium channels in the cell membrane to open. Sodium ions flow through the channels, resulting in a drop in the electrical potential across the membrane. This generates an electrical signal that passes through the nerve and ends up in the brain.

The other type of taste receptors is sensitive to sour and salty tastes; that is to say, especially hydrogen ions (H^+) and sodium ions (Na^+), but also potassium ions (K^+). These receptors are ion channels that cut across the cell membrane. A change in the concentration of these ions is registered by the taste receptors, which leads to a change in the electrical potential across the membrane, and an electrical impulse can be sent to the brain.

THE INTERPLAY BETWEEN SWEET AND BITTER
In contrast to umami, which in its pure form is brought out by only a small number of substances, an incredible range of very diverse substances brings out sweetness and bitterness. For a long time, it was thought that the receptors for sweet and bitter tastes were very closely related. This is due to the fact that many sweet substances, such as the artificial sweetener saccharine, can have a bitter aftertaste. In addition, only very minor chemical modifications can change a molecule from

sweet tasting to bitter tasting. Mirror images of molecules can actually have sweet and bitter tastes, respectively. For example, the artificial sweetener aspartame, which is made up of two amino acids, tastes sweet if the amino acids are left turning. But if the same molecule contains right-turning amino acids instead, it will have a bitter taste.

We have still not identified the mechanism that is responsible for this difference. It is possible that it is due to an as-yet-undiscovered interplay between the taste cells and the presynaptic cells.

Schematic illustration of a taste receptor that is embedded in the membrane of a sensory cell. The membrane consists of a double layer of lipid molecules, which form an effective barrier between the inside and the outside of the cell. The receptor is a large protein molecule that traverses the membrane in serpentine fashion with, in this case, seven twists and turns.

> **TASTE RECEPTORS: THIS IS HOW THEY WORK**
>
> Research carried out recently has gradually reduced the number of different types of taste receptors to a relatively small number. In the case of sweet, bitter, and umami, the main focus of attention at the moment is on two classes, labeled T1R and T2R, which can function independently or in particular combinations. The receptors are all G-protein-coupled receptors. Their defining characteristic is a large terminal domain, which protrudes from the outer side of the membrane. This domain is where taste molecules are identified and bound.
>
> When the taste molecule is bound to that part of the receptor that protrudes out of the membrane (pointing upward in the illustration), what is known as a G-protein is bound simultaneously on the end of the receptor that bends around to the inside of the cell. This binding action sets in motion a cascade of signals, which results in the transmission of an electrical signal through the nerve cells to the brain.
>
> The T1R receptors are distantly related to the receptors that are sensitive to the types of amino acids that function as neurotransmitter substances in the brain.
>
> Surprisingly, we know unbelievably little about the G-proteins that are bound to the receptors on the inside of the membrane. This is why we have only very limited knowledge about what happens at this end point inside the cell and the series of signals that causes the ion channels to open and, in so doing, send the resulting electrical sensory signal to the brain. A single G-protein, referred to as gustducin, has been isolated and shown to couple to the T1R and T2R receptors, and hence is involved in the sensing of sweet, bitter, and umami tastes.

Each class of receptors can have a number of different members, such as T1R1, T1R2, and T2R3, which all belong to the T1R family. The different members can be expressed in varying quantities in the individual taste cells. For instance, some cells may have only T1R3 receptors, while others may contain the combination T1R1/T1R2, and still others may contain T1R1/T1R3.

Sweet, bitter, and umami tastes make use of a variety of receptor combinations. This is in contrast to sourness and also, presumably, saltiness, each of which is based on a single receptor. It is therefore probable that the signal pathways for sweet, bitter, and umami tastes are different from those for the sour and the salty.

SWEET

Given that our gustatory sense perceives a large number of really diverse substances as being sweet, one might think that there would be a correspondingly broad range of taste receptors for sweetness. But in recent years, it has finally been established that the combination T1R1/T1R2 is the most important, and possibly the only, receptor for sweetness in mammals. Its perception of sweet tastes covers everything from naturally sweet substances to artificial sweeteners, sweet amino acids, and certain sweet proteins. The reason that the receptor is sensitive to a range of substances that are, chemically speaking, very different is that the various substances bind to different sites on the rather large part of the protein that protrudes from the membrane.

The detailed molecular structure of the receptor can vary significantly from one mammal to another. For example, mice, unlike humans, cannot taste the artificial sweetener aspartame, and cats are missing the gene that allows them to form T1R2 and, therefore, cannot taste sweetness at all.

As we will see later, the receptors for sweetness are closely related to the receptors for umami taste.

BITTER

Unlike sweetness, for which there is only one type of receptor, there are many receptors that are activated by bitterness. Bitter substances stimulate a large family of about thirty different members of the T2R receptor class. It has now been shown that members of this family are both necessary and sufficient to allow the taste cells to respond to a bitter taste. It is possible that the active T2R receptors are arranged in pairs. Each type of T2R can bind several different types of bitter substances. T2R receptors have a much smaller terminal domain that protrudes from the membrane than that of T1R.

Perceptions of bitterness have to be very unambiguous and wide-ranging, as they are an indication that the food might be poisonous. On the other hand, a bitter taste does not have to be especially nuanced in order to play an important evolutionary role. Consequently, the receptors for bitterness are stimulated by many different chemical substances that are not even remotely related. It has also been discovered that most T2R receptors in the taste buds are bundled together in the same cells, where they act as a very broad-spectrum sensor for bitterness.

SOUR

Over time, many mechanisms and receptor models have been proposed to explain sourness. In all cases, these have involved two classes of membrane proteins—namely, ion channels and membrane pumps—to facilitate the transport of sodium ions, potassium ions, and hydrogen ions across the membrane.

At present, the spotlight is on one particular ion channel, known as PKD2L1, as a receptor for sourness. This ion channel is found primarily in the taste cells that are not sensitive to sweet, umami, and bitter tastes. These taste cells, therefore, act as the sensors for sourness.

The taste cells sensitive to sourness have also been found to host a special receptor, called Car4, that is sensitive to carbon dioxide. Hence carbon dioxide, for example as found in carbonated beverages, stimulates the sour-sensing cells and induces a mild sour taste sensation, although the fizzing and tingling are of a mechanical nature.

SALT

It may come as a surprise that the mechanism behind the perception of saltiness is the one about which we know the least. In 2010 in an experiment with mice, an epithelial sodium channel called ENaC, localized in a specific population of the taste cells, was found to mediate the salty taste.

WHEN WORDS FAIL US: DESCRIPTIONS OF TASTES

Everyday vocabulary can easily come up short when one is trying to describe a particular taste. Because many food cultures have evolved their own individual ways of characterizing a basic taste with a single word, cultural differences can lead to difficulties in describing taste impressions that are less common or unknown in a particular cuisine. For example, people accustomed to food prepared in the Western world are usually able, without hesitation, to categorize substances as having sour, sweet, salty, and bitter tastes. Few would doubt which word to choose to describe definitively the taste of a food such as a lemon. On the other hand, it would immediately become much trickier for them to describe the taste of MSG.

As shown in a classic psychophysical experiment, carried out by Michael O'Mahony and Rie Ishii from the Department of Food Science and Technology at the University of California, Davis, USA, it boils down to a question of expressibility or codability. These researchers compared the taste-naming strategies of two groups: monolingual Japanese speakers and monolingual American English speakers. For the most part, neither language group had much difficulty in choosing a single word to describe taste samples that were sour, sweet, salty, and bitter. But when it came to MSG, there was a significant difference.

The English speakers were able to differentiate the taste of MSG from that of the other taste samples, but they had no single expression to describe it. In fact, they used expressions that bore little resemblance to each other and that would not be classified as basic tastes. They described it as 'salty,' while noting that it was not the same as ordinary saltiness, but they also used descriptors including 'indefinite,' 'fishy,' 'beef bouillon,' and so on. The majority of Japanese, however, linked the taste of MSG to the word umami or expressions closely related to umami, such as dashi (a soup stock), although several of them also said 'salt-like.' Many used the word *Ajinomoto*, which is the trade name for a common MSG taste enhancer. Those Japanese subjects participating in the experiment who were professional tasters tended to employ the more scientific term umami. The researchers concluded that the differences between how the two experimental groups described tastes were not attributable to physiological mechanisms or underlying sensory concepts. Culture, rather than language, was the determining factor in the number of basic tastes that were clearly associated with a single word.

▸ Examples of raw products that most people would, without any hesitation, associate with one of the four classic basic tastes. Sour: red currants with crème fraîche. Sweet: melon with honey. Salty: oysters and sea asparagus (sea beans). Bitter: radicchio and walnuts.

*There is still another quality,
which is quite distinct from all these
[sweet, sour, bitter, briny] and which must
be considered primary, because it cannot be
produced by any combination of other qualities....
It is usually so faint and overshadowed by other
stronger tastes that it is often difficult
to recognize it unless the attention is
specially directed toward it....
For this taste quality the name
'glutamic taste' [umami]
is proposed.*

これらの味とは全く別の、そして他の味を如何に組合わせても作ることができないことから、本源的とみなければならない別の味がある。その味は通常非常に弱く、他の強い味によってボカされるので特に注意をそれに向けないと識別することがむずかしい。私はこの味に「グルタミン酸の味」という名称をつけようと思う。

Kikunae Ikeda (1864–1936)

The fifth taste: What is umami?

Even though the word *umai* had been used in Japan for hundreds of years to signify the concept of something delicious, people only became truly conscious of it thanks to the efforts of a single individual—Japanese professor and chemist Kikunae Ikeda (1864–1936), who set himself the challenge of identifying the substance in Japanese soups that was responsible for their fantastically good taste. He found the answer in 1908.

SCIENCE, SOUP, AND THE SEARCH FOR THE FIFTH TASTE
Japanese soups are based on a stock called *dashi*, which is a very simple, clear broth. It is made by extracting the taste substances from a particular species of brown macroalgae, konbu, and flakes of fish that have been cooked, salted, dried, fermented, and smoked, known as *katsuobushi*. Dashi is rich in umami and is a ubiquitous, indispensable element that is central to all traditional Japanese cuisine.

Professor Ikeda's hypothesis was that the soup must contain a substance that imparted a taste that could not be explained away as a combination of the four common basic tastes. So he set to work on the very labor-intensive, and at times probably very tedious, process of making a chemical analysis of all the ingredients found in an aqueous extract of konbu (*Saccharina japonica*), a species of kelp. He started with 12 kilograms of the seaweed, basically working on his own with a single laboratory assistant.

First, Ikeda discovered that the extract from the seaweed contained an abundance of a particular carbohydrate, mannitol, a sugar alcohol present in the slime that is exuded onto the surface of many brown algae. Mannitol tastes somewhat sweet. He also noted the presence of a number of inorganic salts. Finally, he isolated a component that he identified as a salt of

In 1908, Professor Kikunae Ikeda (1864–1936), a Japanese chemist, discovered that monosodium glutamate (MSG) is the substance that is responsible for the delicious taste of the soup stock dashi. He was the first person to investigate this taste in a scientific manner and introduced the term by which it is now known: umami.

an organic acid, glutamic acid, which is an amino acid. This salt was monosodium glutamate, or MSG. Glutamic acid had actually been identified scientifically in 1866, but since then no real efforts had been made to undertake further studies of either the acid or its salts as they related to taste.

GLUTAMIC ACID AND GLUTAMATE

Glutamic acid is one of the twenty amino acids that living organisms use to build proteins. Some of these amino acids are characterized as essential, as we cannot produce them in our bodies and therefore depend on getting them from the food that we ingest. Ones that we can produce are characterized as nonessential. Glutamic acid is a nonessential amino acid, and our bodies synthesize about 50 grams of free glutamic acid every day. It plays a major part in the synthesis of other nonessential amino acids in our bodies. From their daily food intake, adults typically ingest 10–20 grams of protein-bound glutamic acid and about 2 grams of free glutamic acid, normally in the form of glutamate.

The salts that can be formed from glutamic acid are called glutamates. The most common of these is a sodium salt, monosodium glutamate or MSG, but glutamic acid also forms salts with other compounds such as potassium, calcium, ammonium, and magnesium, all of which are found naturally in foodstuffs. All glutamates bring out the umami taste, but MSG is especially effective because it interacts with another important salt in our diet—table salt (sodium chloride, NaCl).

An adult weighing 70 kilograms has about 1.6 kilograms of glutamic acid in his or her body, most of which is protein-bound. But at any given time there are about 12 grams in the form of free glutamate. The bulk of the free glutamate is found in the muscles (7 grams) and brain (2.6 grams), whereas the blood contains only a very small amount (0.045 grams).

Ingested glutamic acid and glutamate are absorbed in the stomach and intestines via an active transport system located in the cells of their mucous membranes where up to 95 percent of these two substances are metabolized and converted to energy. Hence, only a little glutamate reaches the bloodstream. In fact, glutamate is actually the most important single source for the production of energy during the digestive process. This conversion depends on the concentration of sodium, which is a function of the salt content of the food. Any surplus glutamate is broken down in the liver.

> Apart from being an important component in the makeup of proteins, glutamate is the most widely distributed signaling substance in the human nervous system, as well as the nervous system of all vertebrates, where it functions as a neurotransmitter. A nerve impulse triggers the release of the glutamate from a nerve cell, transporting it to an opposing cell, where it binds to a glutamate receptor, thereby activating the latter. Glutamate plays an especially important role in maintaining brain plasticity; that is, the constant formation, undoing, and re-creation of connections between the nerve cells of the brain on which memory and learning are based.
>
> One could legitimately ask about the extent to which the delicate signaling systems in the brain are affected by the glutamate naturally derived from our food and the additional amounts that we might eat in the form of MSG. Fortunately, the brain and the nervous system are well protected. As already noted, glutamate is metabolized quickly and effectively in the digestive system at the rate of 5–10 grams an hour, and the bloodstream is normally not affected very much by the glutamate content of the food. At least 5 grams of pure MSG must be eaten all at once to bring about a noticeable change in the glutamate level in the blood. The excess tapers off in the space of a couple of hours as the glutamate is gradually broken down in the liver. Furthermore, it is very difficult for glutamate to cross the blood-brain barrier on its own. Nor is that desirable, since the nerve cells in the brain reuse or produce the glutamate they require in order to carry out their signaling functions. This combination of mechanisms allows the body, under normal circumstances, to maintain a delicate balance of neither too much nor too little glutamate.

Professor Ikeda's original 12 kilograms of seaweed had yielded a mere 30 grams of glutamate. From this extracted glutamate, he was able to produce small crystals, which he compared to grains of sand. When he tasted the crystals, he found that they tasted somewhat like a sour version of dashi, but as the sour taste gradually faded, what remained was a taste very much like that of dashi. As a temporary measure and for the sake of convenience, Ikeda suggested that the taste should be called umami until a better name could be found. This never came to pass and the word has been used ever since. He remarked that glutamate had a very distinctive taste. And even though by itself it was not very palatable, it had the effect of making other food taste delicious.

> ### WHAT IS THE MEANING OF THE WORD *UMAMI*?
>
> Ikeda gave a lot of thought to his suggestion of the expression umami to describe the taste of glutamate. Umami (旨味 or うまみ) is derived from the Japanese adjective *umai* (旨い), which can be interpreted in two ways. One has hedonistic connotations, implying that a taste is delicious and pleasant. The second refers to the actual sensory aspect, that something tastes meaty and spicy. As neither of these meanings is specific only to glutamate and both can be applied to other substances as well, *umai* is not the same as the taste of glutamate. In order to differentiate between the two, Ikeda proposed a newly constructed word, umami, which combines *umai* with *mi* (味), which means 'essence,' 'essential nature,' or 'taste.' This is how the taste of glutamate came to be linked inseparably with umami.

Ikeda also compared the intensity of the taste of glutamate with that of table salt and determined that he could perceive the taste of 1 gram of glutamate dissolved in 3 liters of water, whereas about ten times as much table salt was required in order for the taste to be discernible in an equal amount of water. In addition, he observed that the taste of table salt was stronger when glutamate was also present in the solution. Furthermore, he proved that when a solution of table salt and glutamate was diluted, the taste of the salt disappeared first, while that of glutamate persisted even in very weak solutions. Ikeda astutely commented that this relationship between salt and glutamate could be exploited to revolutionize the manufacture of soy sauce, as it has both salty and umami tastes.

Ikeda went on to make another important observation. When an acid, such as vinegar, is added to a glutamate solution, the umami taste is lessened. This occurs because the somewhat stronger acetic acid causes the weaker glutamic acid to be formed from glutamate, thereby preventing the formation of glutamate ions. His conclusion was crystal clear—only the ionized form of glutamate results in umami.

Ikeda published these findings in 1909 in a Tokyo journal in a very short, two-page article written in Japanese. Apart from presenting his discovery of glutamate as the substance that brought out the umami taste in dashi, he also raised a fundamental question about why humans were even able to discern this delicate taste. His hypothesis was that, in the course of evolution, humans developed this ability in order instinctively to choose the food that is most nourishing and that forms the basis for the reproduction of the species. Some plants develop a sweet taste when one chews on them, thereby signaling the presence of energy-rich carbohydrates. Ikeda posited that the same is true for umami; it points to an abundance of nutrient-rich, readily available proteins in the food.

It was difficult for Ikeda's ideas about umami to gain widespread acceptance in the West. As already noted, his original article was written in Japanese; it was translated and published in English only in 2002, after the first umami receptors had been discovered. In addition, because the taste of glutamate is mild and subtle and is often displaced by other taste sensations, particularly salty and sour ones, it was difficult to convince people that umami was a true basic taste. Serious research efforts on umami were not set in motion until the 1980s, and these endeavors gained full academic respectability only after the identification of specialized umami receptors in 2000 and 2002. We will return to this topic later.

FROM LABORATORY TO MASS PRODUCTION
In Japan, Professor Ikeda is regarded as one of the country's ten most distinguished inventors. This is not actually because of his identification of glutamate as the basis of umami, but rather because he immediately grasped its technological and commercial potentials. In so doing, he laid the groundwork for what was to become one of the world's largest multinational industrial enterprises within the food sector, namely, the creation of the company Ajinomoto.

Despite its brevity, Ikeda's original article from 1909 incorporated many of the elements required for the successful technology transfer of his discovery. He noted that it would be expensive to extract pure MSG from seaweeds, but this was probably not necessary if its end-use was as a flavor enhancer. Here it would often be combined with other ingredients, which would not spoil the taste of the glutamate and might even add to the nutritional value of the product. Ikeda also wrote that it should be possible to use industrial hydrolysis to mass-produce glutamate from plant matter, which is rich in proteins. This process breaks down the proteins, releasing their glutamic acid and glutamate contents. Ikeda modestly disclosed in the article that he had already taken out a patent on his discovery one year earlier.

Ikeda's patent for the extraction of glutamate was the cornerstone of the partnership he formed in 1908 with the entrepreneur Saburosuke Suzuki. They set up a company, calling it Ajinomoto, to make MSG for use as a food additive. Right from the start, the product was a great

THE ROLE OF PHYSICAL CHEMISTRY
It is highly likely that Kikunae Ikeda was strongly influenced by important advances in scientific thought that were current around the turn of the twentieth century. Between 1899 and 1901, he had studied in Germany under the legendary chemist Friedrich Wilhelm Ostwald, one of the founders of modern physical chemistry, who was awarded a Nobel Prize a few years later. Ikeda's approach to identifying precisely those ingredients in dashi that were the sources of its umami taste is an excellent example of the application of quantitative physical methods to chemical investigations, which was the hallmark of this new field.

success. Today Ajinomoto makes almost two million tons of MSG annually, which represents one-third of the total world output of glutamate. This works out to an average daily consumption of about half a gram of MSG by every man, woman, and child on Earth.

Ajinomoto means 'the essence of taste.' With yearly revenues of about US$9 billion and subsidiaries located in many parts of the world, the enterprise is one of the jewels of Japanese industry. Its flagship product is MSG, either on its own or in the extensive range of products in which it is combined with other flavor enhancers and spices. MSG is often mixed with salt, for example, 10–20 percent table salt in a typical product. This has the effect of setting a natural limit on the amount of MSG that one is likely to add to the food.

HOW MSG IS MADE

Kikunae Ikeda's original patent described the production of MSG by hydrolysis of wheat protein, about 25 percent of which is glutamic acid. Hydrolysis breaks the proteins down into their constituent amino acids. Since the mid-1960s a cheaper method has been employed, wherein MSG is made by fermenting a number of starch-containing plant by-products derived from corn, sugar beets, and sugarcane (in the latter case, especially in the form of syrup and molasses). In a certain sense, this fermentation process is just as natural as the one that turns milk into yogurt and cheese, soybeans into soy sauce and miso, or cereals and grapes into beer and wine, respectively.

Early glutamate production in Japan was carried out in large ceramic crocks, where wheat gluten was hydrolyzed with the help of a strong acid. Over a period of 20 hours, the ingredients in the crocks were stirred at regular intervals using a long pole.

Currently, the most common method for producing MSG in any part of the world employs a variety of types of yeast and bacteria and involves a myriad of different enzymes. By varying the selection of microorganisms and conditions governing the reactions, it is possible to optimize the production of a certain amino acid—in this case, glutamic acid. For example, the microorganisms *Micrococcus glutamicus*, almost as their name implies, are good at producing glutamic acid. These bacteria thrive in an oxygen-poor growing medium of nutrients and minerals, where they synthesize and excrete glutamic acid onto their surfaces. The acid is released into the medium, from which it can be separated and converted to MSG. No one has as yet discovered why the bacteria make so much glutamic acid.

MSG MOLECULES HAVE TO TURN CORRECTLY

Amino acids and, consequently, also glutamic acid are what is known as chiral molecules, which is to say that they come in two variations that are mirror images of each other, one turning in a counterclockwise direction (signified with an L) and the other in a clockwise direction (signified with a D).

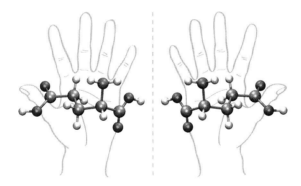

Glutamic acid comes in two configurations, which are mirror images of each other in the same way as the left and the right hands. The salts of the left-turning one (L-glutamate) impart umami taste, while those of the right-turning one (D-glutamate) have no effect on taste.

Only the left-turning molecule, L-glutamate, fits into the glutamate receptor, and it is, therefore, the only one used as a taste enhancer. D-glutamate has no characteristic taste and does not impart umami. Commercially produced MSG is very pure, being made up of more than 99.5 percent L-glutamate. This is considerably purer than the naturally occurring free glutamate in foodstuffs that have umami. It is also purer than that in fermented products, such as soy sauce, which may contain up to 5 percent of the right-turning variant. For this reason, it is highly improbable that insignificant impurities in industrially produced glutamate could lead to health problems, as has been claimed by some skeptics.

GLUTAMIC ACID OR GLUTAMATE SALT?

Glutamate in the form of MSG is actually a salt derived from glutamic acid, and it dissolves more readily in water than the acid does. Once dissolved in water, MSG decomposes into sodium ions and glutamate ions. The amount of glutamic acid formed is totally dependent on the acidity of the solution. The greater the acidity of the solution, the more glutamic acid there is. And as it is glutamate that imparts umami, the taste intensity will decrease as acidity increases. The taste of the glutamate ion will, to a certain extent, also be affected by the sodium ion in MSG, which tastes salty.

MSG is very stable and does not break down in the course of normal food preparation. But when MSG crystals come in contact with water, they split apart into sodium ions and glutamate ions, taking on the form in which they can bring out the umami taste.

The problem with MSG is that it resembles what it is, namely, a white powder or granulate—a chemical, artificially produced in a factory. Just using the words *chemical* and *artificial* is enough to cause many people to balk at the idea of putting such an additive on their food, even though this is not rational. Ordinary table salt and refined sugar (sucrose) are also chemicals, which incidentally have a close physical resemblance to Ajinomoto's glutamate products.

Nevertheless, the only real problem with MSG is that it can be misused. The unfortunate fact is that it is often called on to compensate for loss of taste in prepared foods due to the use of ingredients of poor quality or the way in which they are processed.

▸ A blade of the large Japanese brown alga konbu, on which can be seen white blotches of sea salt, mannitol, and glutamate that have been exuded onto its surface. A small pile of white crystals of pure MSG has been placed on it.

After Ikeda's discovery, the use of MSG in food preparation was an overnight success in Japan and China, whereas it took right up until the 1960s before the practice became widespread in the Western world. In the West it was especially the growing Chinese restaurant sector that adopted the use of MSG, particularly in the United States.

CHINESE-RESTAURANT SYNDROME

To the Editor: For several years since I have been in this country, I have experienced a strange syndrome whenever I have eaten out in a Chinese restaurant, especially one that served Northern Chinese food. The syndrome, which usually begins 15 to 20 minutes after I have eaten the first dish, lasts for about two hours, without any hangover effect. The

A little letter with a huge impact: The 'Chinese restaurant syndrome'

In 1968, *The New England Journal of Medicine* published a short communication, which was not a scientific article but a letter to the editor, from Dr. Robert Ho Man Kwok. In this little letter, which is only about forty lines long, Dr. Kwok describes his personal observations about what he characterized as a "strange syndrome" that he and others experienced after eating at Chinese restaurants. The syndrome resembled a mild version of the hypersensitivity some individuals have to the acetylsalicylic acid found in aspirin. The somewhat vaguely defined symptoms associated with the syndrome included numbness at the back of the neck that gradually radiated to the arms and the back, general weakness, heart palpitations, and thirst. They set in 15 to 20 minutes after the meal had started and lasted for a few hours, but they had no lasting effect.

Dr. Kwok writes that the underlying cause of this reaction is obscure and speculates on a range of different possibilities. He specifically mentions a number of ingredients that were all, at that time, used liberally in Chinese restaurants: soy sauce, cooking wine, MSG, and salt. Any of these in excess could trigger a reaction, which might be more acute when several of them are combined. He is very cautious in his allegations and suggests that it might possibly be that the sodium from the large quantities of table salt, together with that from the MSG, creates an imbalance between potassium and sodium in the body. At the conclusion of the letter, he makes a plea for more information on the subject.

It is noteworthy that the expression 'Chinese restaurant syndrome' and the more recently adopted 'MSG-symptom complex' are not to be found in Dr. Kwok's letter. Rather, it was used as a heading for the letter and was chosen by the editor.

This short letter was to have a huge impact, which is felt even now.

Dr. Kwok's letter attracted much attention and caused an uproar, not least because government authorities had declared that MSG was safe for human consumption. MSG was vocally proclaimed to be a fiendish invention of the food manufacturers and its consequences cited as an example of the dangers of using chemicals in foods. Experiments had shown that MSG in very large doses could cause brain damage in mice. Mothers were panic-stricken when they heard claims that MSG caused lesions in children's brains.

The outcry set in motion a whole series of rigorously conducted scientific investigations, which ultimately did not come up with any evidence that MSG had negative effects. The conclusions drawn were that MSG is a harmless additive that has no adverse side effects on the vast majority of the population, even in extremely large doses. In the course of the experiments, some negative effects on mice did turn up, but they were linked to doses so high that they would be equivalent to having an adult human ingest more than half a kilogram in one sitting. In such large quantities even ordinary table salt is harmful. Furthermore, research subjects who declared in advance that they were hypersensitive to MSG were found in double-blind trials not to be affected by it. Nevertheless, it was not possible to dismiss the finding that some of the symptoms could result from the ingestion of large doses—more than 3 grams of pure MSG—on an empty stomach and without any accompanying food. On the other hand, the experiments showed that it would be wrong to conclude that ingesting MSG in combination with ordinary Western food would bring on these reactions.

In all probability, MSG is *the* food additive that has been subjected to the most thorough scrutiny of all time. There is nothing to indicate that it is more dangerous than ordinary household staples like table salt, vinegar, and baking powder.

The conclusion to be drawn is that those few individuals, no more than 1–2 percent of the population, who think that they suffer from 'Chinese restaurant syndrome' are not very different from those who experience a variety of allergic reactions to other types of foods, both natural and highly processed.

Many authors have set the problem in context by asking why this supposed illness is not labeled 'Japanese restaurant syndrome' or 'Asian restaurant syndrome,' given that the use of MSG as a food additive is more widespread in Japan than in China. There are no reports of negative health outcomes in Japan as a result of a significant intake of MSG. In fact, Japanese people, whose life expectancy ranks among the highest in the world, are generally very liberal in their use of MSG to prepare food, whether the food is home cooked, prepared in restaurants, or produced industrially.

The recommended daily diet for an adult (weighing 70 kilograms) contains about 50 grams of protein. Of this amount, 15 grams are typically made up of glutamic acid, some of which is in the form of free glutamate. It is, therefore, very difficult to conceive how the addition of 0.5–1 gram of glutamate could pose a health hazard. The running debate about glutamate and MSG in food is an example of a wider problem; namely, that even well-informed people are afraid of 'food with chemical additives.'

It is a little discouraging that so many people take a jaundiced view of MSG as something negative and dangerous, and that they remain unconvinced by scientific evidence. These unfavorable impressions continue to cast a long shadow over the amazing story of the Japanese professor who found umami in his miso soup and launched a multimillion-dollar industry that produces the most umami-intensive substance, glutamate.

Despite what Ajinomoto had been able to achieve in the global market of food additives, there was a fly in the ointment. In 1968, Dr. Robert Ho Man Kwok wrote a letter to the editor of a prestigious medical journal in which he described a strange illness that he had experienced after eating in Chinese restaurants. This came to be known as the 'Chinese restaurant syndrome,' and MSG was fingered as the culprit. All subsequent scientific investigations have shown, however, that MSG is just as harmless as table salt.

THE JAPANESE DISCOVER OTHER UMAMI SUBSTANCES
Already in 1913, one of Professor Ikeda's students, Dr. Shintaro Kodama, made a new discovery related to umami. He found that the substance that brings out umami in *katsuobushi* is a ribonucleotide, inosinate or inosine-5'-monophosphate (IMP), derived from the nucleic acid inosinic acid. His work was based partly on that of the German chemist Justus von Liebig, who had isolated inosinate from beef soup in 1847.

Nothing further happened until 1957, when another Japanese researcher, Dr. Akira Kuninaka, discovered and identified yet another nucleotide with umami taste. In the course of studying the biochemistry involved in the breakdown of the nucleic acids in yeast, Dr. Kuninaka found that a ribonucleotide, guanylate or guanosine-5'-monophosphate (GMP), which is derived from the nucleic acid guanylic acid, is also a source of umami. Subsequently, the presence of guanylate was detected in dried shiitake mushrooms, whose rich umami taste was already well known to both Japanese and Chinese cooks. As in the case of glutamate, these ribonucleotides had been identified scientifically in the course of the nineteenth century, but no one had taken an interest in their effects on taste.

Not content with his discovery of the importance of ribonucleotides for umami, Dr. Kuninaka undertook further research and found, much to his astonishment, that there is a synergistic interaction between glutamate on the one hand and inosinates and guanylates on the other. Small quantities of one substance significantly amplify the taste of the other. He had stumbled on the explanation for the magic properties of dashi, which consists of ingredients that contain glutamate (konbu) and inosinate (*katsuobushi*). This was a truly significant breakthrough in umami research. It also explained why vegetarian soup stock prepared with konbu and shiitake has such a powerful umami taste.

Since 1960, IMP and GMP have been produced industrially and are used as additives in many prepared foods, such as pies, chips, noodles, sausages, soups, and sauce bases, often in conjunction with glutamate to take advantage of the synergy between them. Some products contain 1.5 percent nucleotides and are used as taste enhancers, for example in soups. Others may have 8 percent nucleotides and are used as substitutes for dashi. IMP and GMP are stable in water solutions but can be broken down in acidic ones at high temperatures.

Several other 5'-ribonucleotides with umami taste were found later. These include adenylate or adenosine-5'-monophosphate (AMP) and xanthosinate or xanthosine-5'-monophosphate (XMP). The different ribonucleotides occur in more than one chemical form, and their taste intensity varies greatly. Those nucleotides that contain sulfur are notable for their ability to enhance umami.

Ribonucleotides that impart umami are present in many raw ingredients. Inosinate is found primarily in meat, guanylate in plants and fungi, and adenylate in fish and shellfish. On the other hand, konbu contains neither inosinate nor guanylate.

IMP, GMP, AMP, and XMP frequently occur as disodium salts; that is to say, in compounds with two sodium ions bound to each nucleotide.

IT ALL STARTS WITH MOTHER'S MILK
Human breast milk is rich in free glutamic acid, which makes up more than half of the free amino acids in the milk. It is at least ten times as rich in free glutamate as cow's milk. In addition, mother's milk has a small quantity of inosinate, which enhances its umami taste. Breast milk from chimpanzees has even more free glutamate than that from humans, whereas milk from lower primates has only twice as much glutamate as cow's milk. Apparently there is no correlation between the growth rate of a species and the glutamate content in the breast milk.

Interestingly, amniotic fluid contains glutamate, about 2.2 mg per 100 g, in all likelihood exerting an influence on the taste preferences of the child. There is, consequently, no doubt that a child, both before and after birth, is programmed to seek out glutamate and umami.

Behavioral scientists have studied the facial expressions of newborn babies who are given samples of different tastes. Sour and bitter ones cause them to screw up their faces as a sign that these are unpleasant taste sensations. On the other hand, the babies look relaxed and content when exposed to sweet and umami tastes, for example, from lactose and glutamate, respectively, both found in mother's milk. It is clear that our species instinctively will choose food that is sweet and, consequently, rich in calories or that has umami and is nutritious. The presence of glutamate helps babies to accept breast milk far more readily than that from cows. Researchers have argued that newborns have no preference for MSG as such, but rather MSG in combination with a complex mixture of other substances. So it would seem that we are born with a preference for umami.

Taste preference during childhood is, however, very dependent on country of residence and age. Some recent cohort studies indicate that the preference for umami in European children peaks between the ages of six and nine.

Oddly enough, since 1969 glutamate has not been added to infant formula, presumably as a historical outcome of the emotionally laden discussions about the putative negative effects of MSG, which started in 1968. It is paradoxical that infant formula has been robbed of precisely that taste substance that has the most influence on the formation of a child's taste preferences.

UMAMI AS A GLOBAL PRESENCE
Food with umami is not found only in Japanese and other Asian cuisines. The taste is a feature of all culinary traditions, and there are some very good reasons for this. First, everything edible contains components, in the form of proteins and free amino acids, that are potential sources of umami. In addition, the association of umami with deliciousness is a universal trait based on the human taste experience. Food cultures in all parts of the world have, in a sense, been created around efforts to enhance umami taste. We like and find pleasure in food with umami, an aspect of our character that has developed further over the course of human evolution.

But the fact of the matter is that in many languages there is no single expression that denotes this taste sensation. Furthermore, in Western cuisines, in contrast to that of Japan, there is no single, freestanding ingredient that is commonly used by everyone to impart umami to food. In Japan this is dashi, which is inseparably bound up with the fifth taste. In a cuisine that uses very little fat, dashi is an absolutely essential source of savoriness.

▸ Dried shiitake mushrooms, which are rich in guanylate.

It has taken researchers in the Western world a long time to accept that there are taste sensations that cannot be described as combinations of sour, sweet, salty, and bitter. Perhaps this is a reflection of our having been schooled in a rational way of thinking that fosters skepticism until we are faced with scientific evidence. We could not be convinced unless, and until, a sensory-physiological explanation for umami in the form of a molecule (namely, a receptor in the taste buds on the tongue) that clearly spells out umami could be found. Hence, the discovery of the first specialized umami receptor, to be described shortly, was a key development. In parallel with these scientific advances, awareness of the 'fifth taste' gradually slipped into the culinary mainstream as many leading chefs in the West started to embrace umami, partially as a result of the growing interest in molecular gastronomy.

UMAMI HAS WON ACCEPTANCE AS A DISTINCT TASTE
The concept of umami is much broader than the taste of glutamate. Ikeda was well aware of this, and he had noted that the taste of pure MSG was different from that of the naturally occurring umami that was a feature of Japanese dashi. Just the same, MSG proved to be the key to identifying umami as a true basic taste in the sensory-physiological sense. But the work progressed slowly, partly because it proved to be difficult to use animal models for gustatory experiments. Different species, and even subspecies (for example, different types of mice), do not respond uniformly to the substances that bring out umami.

Even though, using sensory-physiological methods, it had already been demonstrated in the 1960s that the taste of glutamate was not dependent on the sour, sweet, salty, and bitter tastes, the big breakthrough in umami research would not take place until almost ninety years after Ikeda had first identified MSG in dashi. In 2000, a group of researchers from the University of Miami in the United States found a receptor in the taste buds of rats that is sensitive to L-glutamate. This receptor, which is distantly related to glutamate receptors in the brain, was given the name *taste*-mGluR4.

The next breakthrough followed rapidly two years later, when two research groups in San Diego, California, in short order found another glutamate receptor. It is a particular combination, T1R1/T1R3, of two different receptors from the T1R family of receptors that are present only in sensory cells.

It is an important distinction that *taste*-mGluR4 is not sensitive to the 5'-ribonucleotides that interact synergistically with glutamate in the umami taste, whereas T1R1/T1R3 is strongly stimulated by the presence of inosinate. It turned out that this observation was precisely the key to deciphering the mysterious mechanism that underlay the incredible synergistic effect on umami that resulted from combining glutamate and 5'-ribonucleotides. This effect had been awaiting a scientific explanation since 1957, when Dr. Akira Kuninaka had made the initial discovery.

Recently, yet another mGlu glutamate receptor, related to *taste*-mGluR4, has been identified. The new receptor is a truncated variant of mGluR1, which is the glutamate receptor in the brain.

So we now know that there are at least three different types of receptors for umami, but we still do not know whether the function of each of them follows the same biochemical signaling pathways to register the taste experience.

AND UMAMI IS STILL CONTROVERSIAL ...
The majority of neuroscientists who work with gustatory perception now seem to be in agreement that there are five basic tastes. But not all of them agree that umami is a true taste in the same sense as the four classical ones.

Hervé This, a French chemistry professor and one of the founders of molecular gastronomy, has pointed out that one of the umami receptors, *taste*-mGluR4, could just as easily be a protein that sends a nerve signal—which is totally distinct from a sensory perception—to the brain. He supports this contention with the fact that molecules other than glutamate activate this receptor. And another observation on which he places great weight is that there is a whole series of different molecules (for example, other amino acids), whose taste also does not seem to be derived from a combination of the four classical basic tastes. Finally, glutamic acid is not the only possible source of umami in Japanese dashi; some also attribute the taste to the presence of alanine, which is also extracted from the konbu.

So Professor This raises the question of whether we should stop thinking in terms of four or five basic tastes, as there may be many more. Actually, according to some neuroscientists, there might be at least fifty different primary tastes.

THE UMAMI INFORMATION CENTER AND 'UMAMI MAMA'
An independent, international nonprofit organization called the Umami Information Center was established in 1982 in Tokyo. Its mission is to publish and disseminate accurate educational, scientific, and culinary information about this basic taste.

The current director of the center is Dr. Kumiko Ninomiya, who is a well-known umami researcher. She has collaborated with chefs and other researchers in all parts of the world regarding their work on the traditional Japanese soup stock dashi, other soup stocks, and a diverse range of umami-rich foods. She has become the very epitome of umami, as is most prominently attested by her nickname, Umami Mama, given to her by the acclaimed celebrity chef Nobu Matsuhisa.

*An attentive
taster will find
something [in] common in the
complicated taste of asparagus,
tomatoes, cheese, and meat, which
is quite peculiar and cannot
be classed under any of the
above-mentioned
qualities
[sweet, sour, bitter, briny].*

注意深く物を味わう人はアスパラガス、トマト、チーズ及び肉の複雑な味の（中）に、共通なしかし全く独特で上記 のどれにも分類できない味を見出すであろう。

Kikunae Ikeda (1864–1936)

1 + 1 = 8: Gustatory synergy

What is very unusual about umami is that the intensity of the taste it imparts is not solely dependent on how much glutamate is present. To a much greater extent, it is affected by synergistic interactions with other substances that increase its gustatory intensity. Most often this involves a 5'-ribonucleotide, especially inosinate and guanylate. Somewhat imprecisely and proverbially, it has been said that the taste imparted by equal amounts of glutamate and a nucleotide is eight times stronger than that produced by glutamate alone. As we will see, however, the synergistic effect can be much stronger.

AMAZING INTERPLAY: BASAL AND SYNERGISTIC UMAMI
In the explanation that follows, we will refer to the two aspects of umami as a basal contribution, based on free glutamate, and a strengthening or synergistic contribution, which is due to the presence of 5'-ribonucleotides.

The fantastic synergy between glutamate and nucleotides can best be illustrated by a simple example. The threshold for being able to taste pure glutamate in water is 0.01–0.03 percent weight/volume (w/v). But if a small amount of inosinate, which has no taste, is also present in the water, the threshold falls to 0.0001 percent w/v, about one hundred times less. (See the tables at the back of the book.) Guanylate is more than twice as effective as inosinate. We also know that there can be similar synergistic interactions between different substances that produce a sweet taste, but they are far less powerful than those associated with umami.

Naturally, cooks in all parts of the world have, throughout the ages, had empirical knowledge of this synergistic effect, but it was not quantified or explained scientifically. It is only recently that we seem to have come closer to an understanding of this mechanism. Paradoxically, the synergistic effect makes it much harder to evaluate the taste of pure inosinate, because human saliva normally contains a miniscule amount of glutamate, about 0.00015 percent.

The culinary arts can be said to be a study of how to maximize umami by taking advantage of the gustatory synergy produced by combining different ingredients. In the preparation of a dish, one will typically

DETECTING UMAMI SYNERGY ON THE TONGUE AND IN THE BRAIN

The Japanese researcher Shizuko Yamaguchi carried out a seminal study based on psychophysical observations, which resulted in an actual formula for the synergy in umami. As shown in the following illustration, the subjective experience of taste intensity increases when glutamate and inosinate are present at the same time. In the experiment, the total concentration of the two substances remains constant (50 mg/100 g), but the proportion of each varies. At this concentration, both pure glutamate and pure inosinate result in very weak taste intensity, but when they are mixed together, the intensity increases and is strongest when there are equal amounts of each. The bell curve derived from these measurements has given rise to the popular saying about synergy in umami: 1 + 1 = 8.

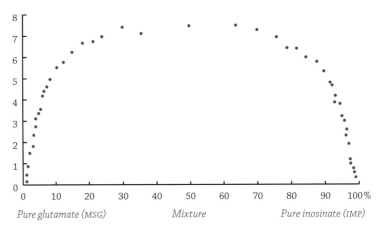

Curve showing the subjective perceptions of the taste intensity of umami produced by a mixture of MSG and IMP, where the proportion of the two components in the mixture varies but the combined amount is constant and lies below the detection threshold of each of the two components. The curve shows that when small quantities of one are added to the other, the taste intensity rises dramatically. Taste intensity is strongest when there are approximately equal amounts of the two components.

> Recent brain scans of research subjects who tasted liquids with umami, in the form of both pure MSG and mixtures of MSG and IMP, support these classical psychophysical measurements. The scans show that there is an area of the brain that responds very actively to umami and that this response is not linear. Measurements of the same type indicate that the area of the brain in question also responds to sweet tastes. Interestingly, the scientists think that this area, part of the orbitofrontal cortex, is the seat of a sort of reward system that is activated when there are advantages to be gained by the individual. This correlates well with the hypothesis that sweetness and umami, from the earliest times, signaled foodstuffs that were rich sources of energy and nutrition, respectively.
>
>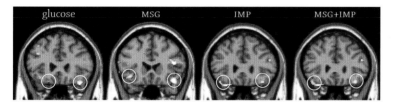
>
> Brain scan of the orbitofrontal cortex of a research subject who is tasting something sweet (glucose) and substances imparting umami (MSG and IMP separately and in combination). The images show that sweetness and umami are registered in the same area of the brain.

incorporate some ingredients that contribute glutamate and others with nucleotides. (See the tables at the back of the book.) Examples of this can be found in just about any cuisine anywhere in the world. But there are none that have a single central element that has put the stamp of umami on an entire food culture in the way that dashi has on Japanese cuisine. As explained a little later in this chapter, this was why we chose to take this versatile soup stock from the kitchen to the laboratory, in an effort to unlock more of the secrets of umami.

JAPANESE DASHI: *THE* TEXTBOOK EXAMPLE OF UMAMI SYNERGY
Dashi is mentioned for the first time in Japanese writings from the eighth century in connection with the fish sauce *katsuo-irori*, an extract of dried bonito (*katsuo*) boiled in water. *Dashi* can be loosely translated as 'cooked extract.' It is a water extract made up of two ingredients, konbu and *katsuobushi*, which contribute glutamate and ribonucleotides, respectively. Dashi is *the* textbook example of umami synergy.

The preparation of dashi starts by placing konbu in cold water. Most recipes stress that the water is then to be warmed up to just under the boiling point, at which time the seaweed is promptly removed. Under no circumstances is the water to boil, because that will result in a more bitter extract. There are, however, indications that having more control over the extraction process will yield a better outcome. Experiments have shown that a temperature of 60–65°C is optimal and that the best results are achieved by allowing the seaweed to steep at this temperature for an hour in a vacuum-sealed bag in a water bath. Some chefs have asserted that higher temperatures break down glutamate, but this is not correct, given that glutamic acid is stable up to at least 150°C.

Freshly shaved flakes of *katsuobushi* are then added to the extract, which starts to give off a wonderful, aromatic smell of smoked fish. Again, the best result is achieved if the temperature is not too high, preferably 60–70°C. Any foam that forms on the surface is skimmed off, as it imparts a disagreeable and slightly astringent taste of oxidized fats. The stock is strained to remove the fish flakes. If the stock is not to be used immediately, it is cooled to prevent the volatile taste substances from evaporating. The newly made dashi keeps in the refrigerator for a couple of days and in the freezer for up to 3 months.

AN EXPERIMENT IN UMAMI SYNERGY
All you need to perform this little kitchen experiment is a can of water-packed tuna and a little tomato paste. The tuna contains inosinate and the tomato has glutamate, which together will create synergy, intensifying the taste of umami.

Drain the tuna and taste it. Rinse your mouth with clean water. Then take a little piece of tuna and crush it with a fork. Add a small amount of tomato paste and mix it in thoroughly. Taste the mixture and note the difference.

Water quality can make an enormous difference in the taste of the dashi, just as it does with tea. Hard water is a less effective medium, whereas soft water will extract more of the salts and the result will be a tastier stock. The hardness of water, which varies from place to place, is determined by its content of calcium and magnesium salts. Soft water, preferably with less than 50 milligrams of calcium oxide per liter, is considered the best for making dashi. In those areas where the tap water supply is quite hard, it can be advantageous to use soft, non-sparkling mineral water instead.

Even though the basic recipe for dashi is very simple, it is still possible to vary it in a number of ways to draw out more, or fewer, harmonious umami tastes. This is because the stock is a water extract that is prepared over a relatively short period of time. This differs from the usual approach in Western cuisines, where meat or soup bones, together with vegetables, are typically simmered for hours, yielding a very robust stock. It is often said that if one puts ten cooks to work making Japanese dashi, they will produce ten different types.

THE ART OF MAKING JAPANESE DASHI

There are many ways of making dashi using a variety of ingredients, but they are all based on extracting glutamate from seaweed (konbu) to obtain basal umami. The outcome depends on the quality and freshness of the raw ingredients, as well as on the care taken in preparing the stock. Different cooks all have their own ideas about steeping time, water quality, temperature, and the shelf life of the stock. The traditional way of making dashi uses konbu and *katsuobushi*. Usually two extracts, first dashi (*ichiban* dashi) and second dashi (*niban* dashi), are made one after the other, reusing the ingredients. Only a portion of the glutamate in the seaweed is extracted into the dashi; the amount depends on the type of konbu, the hardness of the water, and the method used to make the stock. The recipes that follow also incorporate suggestions about alternative ingredients that you can use if you are not able to find konbu and *katsuobushi* or wish to make a purely vegetarian stock.

FIRST DASHI

The first extraction is the finest, yielding the clearest, most delicate dashi. Place the seaweed in water, preferably soft, for about half an hour. Then warm the liquid to a temperature of about 60°C (140°F) and keep it at that temperature for half an hour. It is easiest to do this using a water bath. Remove the seaweed and heat the liquid to 60–70°C (140–160°F), and then add the *katsuobushi* flakes. When the flakes have sunk to the bottom, skim off any froth that has formed on the surface and quickly strain the liquid through a very fine-mesh sieve to remove the fish flakes. The dashi is now ready for use, for example, in a consommé.

1 L (4 ¼ c) soft water • 10 g (⅓ oz) dried konbu • 25 g (1 oz) katsuobushi

SECOND DASHI

The second extraction reuses the seaweed and *katsuobushi*. Water is added to it and it is heated to the boiling point and then allowed to simmer for about 10 minutes. Strain the liquid through a fine-mesh sieve to remove the seaweed and *katsuobushi*. This extract is darker and has a stronger taste than the first one. The higher temperature has drawn out more glutamate, but also bitter substances. For this reason, the second dashi is suitable for heartier soups, such as miso soup.

konbu and katsuobushi from first dashi • ½ L (2⅛ c) water

NIBOSHI DASHI

The *katsuobushi* in first and second dashi can be replaced by small dried fish such as anchovies or sardines, known as *niboshi*, which are a more robust and stronger-tasting source of inosinate. The heads and stomachs of the dried *niboshi* should be removed before use, as they will otherwise impart a bitter taste. Soak konbu and *niboshi* in cold, soft water for about half an hour. Then heat the solution to the boiling point. Skim off any froth that may form on the surface and pour the solution through a fine-mesh sieve to strain out the solid pieces. *Niboshi* dashi is a strong-tasting, robust stock, which is most suitable for miso soup.

KONBU DASHI

If you want to avoid using fish in order to prepare a truly vegetarian dashi, you can make an extraction based solely on konbu. This will yield a dashi with glutamate, but because it lacks the inosinate from fish the umami taste will be less intense and complex.

This simple konbu extract is commonly prepared according to either of these two ways. To make a very clear, light stock, simply soak the seaweed in cold water for about a day. Or you can follow the method with a warm-water extraction as for first dashi, but without adding the *katsuobushi*. The warm water extract is a little darker and has a slightly stronger taste than the one made with cold water.

1 L (4¼ c) soft water • 10 g (⅓ oz) dried konbu

SHŌJIN DASHI

Shōjin dashi is also a vegetarian stock, but it is a true dashi in the sense that it depends on synergy to attain a perfect umami taste. As in all other Japanese dashi, glutamate is obtained from seaweed (konbu), but 5'-ribonucleotides are obtained from guanylate in dried fungi, especially shiitake mushrooms, rather than from inosinate in fish. It is important to use dried mushrooms, as these have a greater concentration of guanylate than fresh ones.

Rinse the mushrooms and soak them in cold water for 5 to 6 hours. To speed up the process, you can shred the mushrooms before soaking them. Strain the resulting extract and reserve it. As for first dashi, soak the seaweed and then warm it to 60°C (140°F) for about half an hour. Remove the seaweed and mix in the extract from the mushrooms. The *shōjin* dashi is now ready for use, for instance, in a consommé with vegetables and tofu.

1 L (4¼ c) soft water • 100 g (3½ oz) dried shiitake mushrooms
1 L (4¼ c) soft water • 10 g (⅓ oz) dried konbu

INSTANT DASHI

If you do not have time to make dashi from scratch, a dry soup powder, e.g., Hon-dashi, consisting of dehydrated dashi containing *katsuobushi*, often with some MSG added to it, can come to the rescue. Packaged preparations for Japanese consommé or miso soup are generally based on dashi powder. Pure konbu dashi powder is also commercially available.

NORDIC DASHI

Once one has realized that it takes two components, namely, glutamate and one or more ribonucleotides, to make a clear dashi with good umami, there is nothing to stop one from improvising with local ingredients. These can be chosen by consulting the tables at the back of the book. And then, armed with knowledge of the synergistic properties of umami, all that remains is to experiment with different combinations and methods of preparation until the desired taste is achieved. It becomes an exciting venture into the realm of molecular gastronomy.

Practitioners of the New Nordic Cuisine have a dogma that all ingredients must be from the region. One can remain faithful to the idea of the traditional Japanese dashi while using local seaweeds, such as sugar kelp, winged kelp, or dulse, as a source of glutamate. The problem, however, is that there is nothing that is truly equivalent to *katsuobushi*. But it could be replaced by dried Nordic mushrooms, cured pork, or prepared chicken. Even though the Nordic varieties of seaweeds have much less free glutamate than konbu, and even though neither mushrooms, pork, nor chicken have as much guanylate or inosinate as *katsuobushi*, experimentation has shown that one can make a reasonably tasty dashi by using another alga such as dulse or winged kelp. The finished stock is mild, with a slightly floral taste.

It is, however, possible to go one step further and create another type of dashi, resembling a vegetarian *shōjin* dashi, using ingredients that are available practically anywhere in the world. It calls for only two things—

Three types of dashi, from top to bottom: konbu dashi, first dashi, and dashi made with the red alga dulse.

Dashi closer to home—a Japanese soup with a Scandinavian twist

The quest for umami and our wish to participate in a pioneering experiment to create a truly Nordic dashi took Ole to the very epicenter of the new northern gastronomy movement—Nordic Food Lab, located in a small gray houseboat anchored in a side canal in the old inner harbor area of Copenhagen. This unusual food laboratory was established as an offshoot of the standard-bearer of the New Nordic Cuisine, Restaurant noma, which is located nearby on the main canal in one of the beautiful, renovated old warehouses of the North Atlantic House complex. This center is a cooperative effort among Denmark, Iceland, Greenland, and the Faroe Islands, dedicated to preserving and promoting their culture and arts. So both the laboratory and the restaurant are in a setting that is Nordic to the core.

Nordic Food Lab provides the framework for a culinary research project that is based on the idea of seeking out deliciousness in Nordic ingredients. It functions as a kitchen in which one tries to combine a chef's intuition and experience with scientific knowledge and methods, using a systematic approach. We had set aside two days to go hunting for a Nordic counterpart to Japanese dashi, a new northern 'mother stock' for imparting umami.

A large kitchen with a panoramic view of Copenhagen's harbor has been installed on the uppermost deck of the houseboat. The large room is filled with light reflecting off the water of the inner harbor, which bustles with the comings and goings of small boats. Even though there are counters, sinks, cooking stations, and a great deal of kitchenware, we should actually refer to this space as a laboratory. In contrast to a normal kitchen, plastic containers, all neatly labeled with descriptions of their contents, are systematically arranged on the work surfaces and stacked in the many refrigerators. This is where research on Nordic food ingredients is carried out. The underlying challenge is to create new dishes, the best of which may end up on the menu at noma. But the laboratory also has more ambitious goals. It wants to inject a breath of fresh air into traditional Nordic cuisine for the benefit of everyone, not just the patrons of an elite restaurant. It hopes to reawaken interest in the special tastes of local products, which are a reflection of the particular climatic and geographical conditions that prevail in these northern areas.

The small kitchen is a hive of activity from morning until night. The working language among the chefs and assistants, who are recruited via an international network, is overwhelmingly English. In fact, the atmosphere is surprisingly like that of a dedicated university research laboratory.

At Nordic Food Lab, one can find chefs who have worked at some of the best and most innovative restaurants in the world, such as The Fat Duck in Great Britain and El Bulli (now closed) in Spain. Others have won major prizes for their culinary skills. They are busily engaged in discussions, making notes and drawings on whiteboards. Those who are less experienced are instructed and mentored by those who are masters. Plans are made and the results of experiments are evaluated on an ongoing basis. Tasting spoons are dipped into the food and noses inhale aromas—technological progress notwithstanding

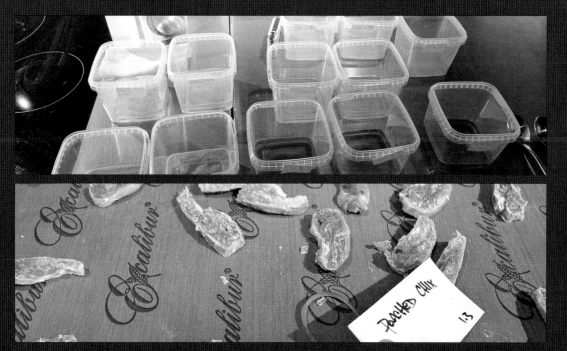

Experimenting with a Nordic version of dashi at Nordic Food Lab in Copenhagen, Denmark.

the most precise measuring instruments in the kitchen are still the human senses of taste and smell. Then there are the incredibly accurate quantitative tools—scales and thermometers and thermostats, which enable the cooks to weigh out fractions of grams and to control with great precision the temperature in pots and water baths.

A whole series of preparations had been made for this set of experiments, which were aimed at creating a new version of dashi from purely Nordic ingredients. Inspired by the Japanese stock, the chefs had gathered ingredients that they hoped would impart sufficient umami. These included seaweeds from cold northern waters, in particular, Danish sugar kelp, Icelandic dulse, and winged kelp from Greenland. Their contribution would be glutamate. The chefs had also prepared several different types of dried chicken, which had first been cooked, smoked, or salted. The chicken was to be a source of inosinate, which works synergistically with the glutamate to yield more umami taste. Another synergistic umami substance, guanylate, could be extracted from the dried button and porcini mushrooms that were on hand. In addition, there was some air-dried smoked ham from the northern tip of Jutland. It could be used to add more glutamate, if the seaweeds turned out to have insufficient quantities.

It appeared that there would be a vast range of possibilities for making dashi. We could use different types of seaweeds and soak them in water at varying temperatures. The extraction could take place over shorter or longer periods of time. Each extraction could then be combined with cooked, smoked, or salted chicken, and the chicken could be cut up into smaller or larger pieces or granulated. Again, the temperature at which this was to take place and the duration of the cooking period could be adjusted. The question then arose as to whether or not to add the fungi to the dashi with the chicken, and whether the mushrooms should be shredded or chopped up first. And for how long and at what temperature should the mushrooms be in the solution? Finally, we also had the possibility of bringing the air-dried ham into the act. There were so many possible variations that it started to look as if working through all of them systematically would become a very time-consuming project.

We had to pare down the list and place our bets on a narrower range of possibilities. So at this point, the expertise and intuition of the chefs took over. We already knew from experience that extracting glutamate from seaweeds should not take place at too high a temperature. In the case of Japanese dashi, the konbu-water solution is traditionally heated to just below the boiling point in order to prevent the release of unwanted bitter substances. We also knew that the taste of dulse changes if it is steeped at over 60°C. So we decided to steep all three types of seaweeds at 60°C for half an hour. This would give us three different extracts as our basic dashi.

A tasting session came next. We quickly eliminated the extract made with sugar kelp. It neither tasted good nor smelled appetizing and, besides, the seaweed had exuded so much slime (actually, polysaccharides) that we could barely get it onto a spoon. Perhaps the blades of kelp that we used were too large or coarse; we had to mull over whether we should try again with some finer, thinner blades or possibly some dried blades that had aged for a while. It turns out that if one ages the sugar kelp for about a year, the taste greatly improves and the result is less slimy.

The dulse extract had a very light color and, in contrast to the extract from sugar kelp, it tasted wonderful. It had delicate floral notes of violets and pansies. Even though the taste was mild, it had an appreciably full mouthfeel. It was unmistakably umami, and there was little doubt that we were onto something that would take us further.

The winged kelp extract had a strong taste of the sea and seaweed. It was pale green and actually tasted rather like a vegetable, something along the lines of cabbage or the stalk of a cauliflower. This might be due to some sulfurous substances that had been drawn out of the seaweed. It was not obvious that it was worth pursuing this option further, but we decided to give it a second chance.

For our next step, we ran two systematic sets of experiments in parallel, using the dulse and the winged kelp stocks. The goal was to enhance umami with the addition of chicken. First we tasted the chicken samples and very quickly decided in favor of the cooked chicken that had been lightly salted, dried, and toasted. But the sample that was only salted and dried was also a possible candidate. We tested them both in the dulse dashi. The pieces of chicken were shredded finely, and 2, 4, and 6 grams, respectively, were added to 250 grams warm dashi, which was kept at 60°C. After they had soaked for 10 minutes, the liquid was put through a sieve. The clear stock was placed in carefully labeled containers and lined up on the counter. Keeping track of all the possible permutations was proving to be a challenge.

We then moved on to another set of taste tests, both when the dashi was warm and when it had cooled. More cooks were called in, and they arrived, armed with their tasting spoons, to sample and discuss the merits of the stocks. There was overwhelming agreement that the dulse dashi was still good and that the taste of umami in the one with the uncooked, but salted and dried, chicken was an improvement over the basic version. It would appear that there was synergy between the substances from the seaweed and those from the chicken.

Much to our surprise, it turned out that the taste of the extract from the winged kelp had changed completely after the addition of the chicken, to the extent that there was now much more umami. At the same time, there was a delicate balance between it and the original interesting sea and vegetable tastes, which had been substantially dampened by the chicken. This was a totally unexpected result that had great potential.

As the first day's work to create a Nordic dashi was coming to an end, we decided to concentrate our next round of experimentation on the two leading candidates—one stock made with dulse and the other with winged kelp, both with the addition of 4 grams of uncooked chicken that had been salted and dried.

The next day, we took another step in the direction of complexity. We would add another synergistic element—namely, guanylate from dried fungi—to the two types of dashi. We quickly discovered that an extract with the dried porcini mushrooms did nothing for our taste buds. The fungi imbued the mixture with a very metallic taste, and the extract had a quite dark color. So we turned to the dried button mushrooms, which we grated into a light powder. We decided to use 1 gram of the powder for 250 grams of seaweed-chicken dashi. We let the ingredients steep together for 10 minutes and immediately realized that we were on the right track. We agreed that the mushrooms had added a new layer of complexity, which we had a hard time describing, but most likely it was umami. The tasting spoons went to work once more, and we invited any chefs passing through the laboratory to sample the result. We were beginning to feel that perhaps we had hit upon a usable dashi.

Of course, we could not leave without finding out whether the air-dried ham might also improve the dashi. But here we were disappointed. Its salty and smoky character had nothing to contribute, and it did not enhance umami. Fortunately, the ham was primarily there to be eaten, and it was whisked away to become part of our lunch. Another experiment that went by the wayside was our attempt to strengthen the umami taste of the dried mushrooms by first soaking them and then lightly toasting them in a pan. This was completely unsuccessful.

Our experimentation was almost at an end, but we thought that perhaps we should just try to fine-tune the salt content of our dashi. It is well known that table salt (sodium chloride) and sodium glutamate have a synergistic effect on the sensitivity of the taste buds toward saltiness. The result is that umami intensifies the taste of ordinary salt. We made a seaweed salt by toasting a little dried granulated winged kelp in a pan and then mixing it with a few tiny flakes of sea salt. The toasted seaweed had a slightly salty and bitter taste, with nutty overtones, and the sea salt added the final touch. It was now ready to be mixed into our dashi to bring it up to the desired degree of saltiness in perfect balance with umami.

The result of our attempts to create a Nordic dashi had produced two types of dashi that tasted good and had been made exclusively from Nordic raw materials. Even though we had not completely achieved our goal, we had shown that it could be done. There was ample reason to think that further experimentation would yield even more delicious results with more umami. All we would need to do is take into account the complexity of finding the optimal relationship in the proportion of the ingredients and the absolutely best methods of preparing them, invite the chefs to exercise their intuition and creativity, and apply them all in a systematic, scientific manner.

water in which potatoes have been boiled and dried porcini or button mushrooms. When unpeeled potatoes are boiled in lightly salted water, glutamate is drawn out. Older potatoes, especially those that have started to sprout, yield a stronger extract. The sprouting process causes the potato proteins to break down, thereby releasing more free glutamate.

To increase the glutamate content of a vegetarian dashi, you can squeeze the juice out of whole sun-ripened tomatoes, seeds, pulp, and all. The resulting stock has only a faint taste of tomato but has a fair degree of sweetness due to the high sugar content.

We can come even closer to the taste of traditional Japanese dashi made from seaweeds and *katsuobushi*. The secret is to use smoked shellfish; we have had good results by using powder made from smoked shrimp heads. When combined with potato water or juice from sun-ripened tomatoes, this powder helps to produce a fantastic dashi with a slightly smoky taste reminiscent of *katsuobushi*. If only potato water and smoked shrimp heads are used, the stock is made entirely from kitchen waste products. ⋯> *Potato water dashi with smoked shrimp heads*

SEAWEEDS ENHANCE THE UMAMI IN FISH
Fish, especially those from saltwater, contain free amino acids, for example, glycine, which tastes somewhat sweet, and glutamic acid, which is a source of umami. In addition, a newly caught, very fresh fish has free nucleotides, especially inosinate and adenylate, which are formed when the cells of the fish produce energy by breaking down adenosine triphosphate. The nucleotides reinforce umami but, unfortunately, these delicious taste substances slowly start to dissipate after the fish has died.

A classical Japanese technique called *kobujime* involves wrapping fresh white fish in seaweed. Before the advent of refrigeration it was used to extend the length of time during which the fish could be kept without spoiling.

While the white fish generally has only about 12 milligrams of glutamate per 100 grams, it does contain inosinate. When a fillet of raw white fish—for example turbot, brill, or flounder—is wrapped in konbu or sea tangle and allowed to cure in the refrigerator for one or two days, glutamate is transferred from the seaweed to the fish, which ends up with more than 300 milligrams of glutamate per 100 grams. The glutamate interacts synergistically with the inosinate to produce a much more pronounced umami taste. This, incidentally, turns the white fish fillet into an ingredient that is eminently suitable for use as sashimi. The seaweed can be reused to make soup.

> #### HOW TO MAKE SMOKED SHRIMP HEADS
>
> Shrimps are a good source of 5'-ribonucleotides, in particular inosinate and adenylate, providing for synergistic umami. The heads are usually discarded but can easily be turned into a versatile source of umami. First, the heads of small shrimps, either raw or lightly cooked, are smoked in a smoker. If you do not have access to a smoker, it is still possible to make them in a pot with a closed lid on the stove. Place some small grilling/smoking wood chips (for example, apple wood) in the bottom of the pot, place the shrimp heads in a strainer that fits completely inside the pot, and smoke at low heat for 10–15 minutes. Then put the smoked shrimp heads in a food dehydrator or spread them out on a baking sheet and dry them in the oven at 70°C (160°F) until they are fully dry. Grind the dried shrimp heads to a powder with a mortar and pestle or grind in a coffee mill. This powder can be stored in a closed container for months.

Smoked shrimp heads.

POTATO WATER DASHI WITH SMOKED SHRIMP HEADS

½ kg (about 1 lb) mature potatoes, unpeeled
1 L (4¼ c) soft water, spring water if available
4 g (¾ tsp) salt
powder made from smoked shrimp heads

1. Scrub the potatoes thoroughly, add the water and salt, and boil them over low heat until they are tender.
2. Pass through a sieve to remove the solid pieces, and reserve the basic stock. Add the powder made from the smoked shrimp heads to taste.

THE FULL-BODIED VERSION

Cooking 100 g (3½ oz) of dried porcini mushrooms or edible field mushrooms along with the potatoes makes the stock even better. To add more umami, combine equal parts of potato water and tomato juice made by squeezing the insides of ripe tomatoes.

1 + 1 = 8: Gustatory synergy

MANY SUBSTANCES INTERACT SYNERGISTICALLY WITH UMAMI

More than forty different chemical substances are now linked to umami, but glutamate is still the one that produces the most intense basal umami taste.

Umami is also ascribed to succinic acid, which is found in both shellfish and sake. The same is true of the amino acid theanine, which is the main ingredient in the taste substances found in tea leaves, where it accounts for half of the free amino acid content. Fermented tea like *kombucha* will have even more umami taste,. There is also a series of small peptides, which are said to draw out or enhance the umami taste.

It turns out that yet another 5'-ribonucleotide, adenylate or adenosine 5'-monophosphate (AMP), which is derived from a nucleic acid, adenylic acid, also interacts synergistically with umami. Adenylate is found in fish and shellfish, for example, and in particularly large quantities in lobster, shrimps, scallops, and squid.

A BREAKTHROUGH DISCOVERY OF YET ANOTHER SYNERGISTIC SUBSTANCE

A few years ago, a group of scientists led by Thomas Hofmann from the University of Münster, Germany, searching for new substances that might impart umami, found a chemical substance called alapyridaine. This substance is not present in the raw ingredients, but it is released as beef is cooked to make a soup stock. It can also be produced artificially in the laboratory by heating a sugar-amino acid mixture. Alapyridaine has a discernible effect on umami and is formed only when the raw ingredients are processed.

Alapyridaine itself has no taste. Quite surprisingly, however, it enhances not only the umami taste, which is due to glutamate, but also strengthens the sweet and salty tastes. On the other hand, it has no effect on bitter and sour substances. So it is actually a substance that can make candy and chocolate taste sweeter without the addition of more sugar. The researchers have found that alapyridaine also interacts synergistically with guanylate, resulting in another potential way to increase umami.

Alapyridaine is one of the first substances to be discovered that can enhance more than one of the five basic tastes. It is likely that many similar substances will be found in due course.

THE INTERPLAY BETWEEN GLUTAMATE AND THE FOUR CLASSIC TASTES

Monosodium glutamate has an effect on how we experience sour, sweet, salty, and bitter tastes, but the relationship is complex. The most dominant one is the way it harmonizes with salt, which so often determines whether the food has much taste at all.

Table salt is NaCl, which means that, like MSG, it contains sodium. When these two substances are dissolved in water, their sodium takes on the form of sodium ions (Na$^+$). Taste tests have been carried out in which the sodium concentration from table salt and MSG together was kept constant in a food, but the relative proportion of each was varied. In all cases, the research subjects reported that decreasing the salt content while increasing that of the MSG made the food taste saltier and improved its palatability; there was, consequently, no reason to add more salt. The experiment demonstrated that the judicious use of MSG makes it possible to decrease the salt content of food without losing even a little of its salty taste. This is good news for people with high blood pressure, a health problem that is on the upswing in a large proportion of the population.

One can draw the conclusion that table salt interacts synergistically with glutamate but does not have the same effect on taste as MSG, even though the latter can have the effect of decreasing the amount of salt needed to make the food palatable. Furthermore, the actual taste threshold for salt is not changed by the presence of MSG.

MSG cannot be used as a replacement for sugar, but it seems to be able to intensify the taste of small amounts of sugar. Consequently, MSG can also moderate sour tastes in such products as sweet pickles, salad dressing, ketchup, tomato juice, and other foods with tomatoes, even though the actual taste threshold for sourness appears to be reduced slightly by the addition of MSG. On the other hand, the taste threshold for sweetness does not seem to change when MSG is also present.

The taste threshold for bitterness is the one on which MSG has the greatest impact. In some instances, it has been found that MSG can lower the taste threshold for detecting bitterness by up to thirty times. Still, MSG can be used to mask a bitter taste, because the intensity of bitter is diminished in the presence of umami.

> **A SIMPLE TASTE TEST: UMAMI VS. SALT**
>
> To find out more about the interplay between salt and umami while at the same time gaining experience in recognizing umami, you can carry out this simple kitchen experiment suggested by Anna and David Kasabian.
>
> Dissolve ¼ tsp table salt in 200 mL (⅘ c) demineralized water and let it come to room temperature. Divide the solution into two equal portions and add a pinch of glutamate to one of them. Taste the water that has the salt-only mixture and note the intensity of the saltiness. Then rinse your mouth thoroughly with clean, room-temperature water, repeat the taste test, and rinse again. Now taste the water with both salt and glutamate in the other container. Note that two things have changed. The salty taste is much stronger, and another taste sensation has put in an appearance. That new taste is umami.
>
> **UMAMI CAN TAKE THE EDGE OFF A BITTER TASTE**
>
> Another little kitchen experiment suggested by Anna and David Kasabian will show you how you can mask a bitter taste with umami.
>
> Coffee tastes bitter because it contains bitter substances such as caffeine. Let a cup of coffee cool to room temperature and taste it thoroughly, swirling it around in your mouth. Make a note of how bitter it tastes. Add a tiny pinch of glutamate to the coffee and ensure that it dissolves completely. Taste the coffee again and judge for yourself whether it is less or more bitter than it was at the beginning.

As a concrete example, let us examine one way in which umami interacts with bitterness. Liver from both fish and poultry has a bitter taste that can have a lingering and somewhat unpleasant aftertaste. By soaking the liver in water or milk, one can remove some of the bitter taste. But it is still important to retain some of it in order not to erase completely the characteristic taste of the liver. This amounts to a challenge to cooks to find a way for diners to encounter the bitterness in such a way that it is carefully balanced by the other taste impressions. It is possible to soften the liver's bitter taste by introducing umami—for example, by adding tomatoes or anchovies to the recipe—and in that way attaining a balance between the two. An excellent example of this is the preparation of monkfish liver, which itself has some umami. ⇢ *Monkfish liver au gratin with crabmeat and vegetables (page 58)*

Umami-rich 'foie gras from the sea'

Only a few decades ago, monkfish was disdained as a relatively uninteresting bycatch when trawling for groundfish or dredging for scallops. There is little question that its appearance worked against it, as it was considered to be one of the ugliest creatures in the sea. With its huge head and enormous mouth, a peculiar sort of antenna used to lure prey toward it, and clumsy mitten-like dorsal fins, it was unlikely to make an appearance on any dinner table as a grilled whole fish.

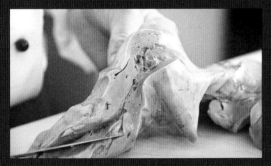

Monkfish liver.

In the past few years, though, monkfish has come to be appreciated for its taste, and it is now such a sought-after delicacy that there is no longer any justification for calling it 'poor man's lobster.' The flesh from the tail and the cheeks costs a small fortune at the fish store. But there is yet another delicious part of the fish, namely its liver, which has not made many inroads in Western cuisine. This is puzzling, as monkfish liver is held in high esteem in Japan, and some, who have come to know it elsewhere, call it 'foie gras from the sea,' putting it on a par with goose liver.

Monkfish liver certainly deserves to take a place alongside the much more common cod liver, which is already a familiar item in some cuisines. It can be prepared so that it is soft and creamy, with good mouthfeel. It has tastes of the sea, but these are mingled with its fattiness and inherent slightly sweet and partly bitter and nutty tastes. In contrast to liver from poultry, or for that matter from cod, which is brownish, it is pale orange, providing a beautiful color contrast in a dish.

A fatty goose or duck liver can have a fat content of 50–60 percent. The fat, which is overwhelmingly made up of unsaturated fatty acids, is distributed throughout in small droplets, which results in a pleasing, delicate mouthfeel. Monkfish liver has only about half as much fat as goose liver, but it is also primarily made up of unsaturated fatty acids. The liver of a fully grown monkfish can be quite large, weighing as much as half a kilogram. Small livers from younger fish are much finer than the large ones, which have a more bitter taste and a coarser texture. Monkfish liver contains only about one tenth as much iron as goose liver, and it is consequently less characterized by the taste of iron than poultry liver is. But it is rich in umami, which can be enhanced by the way it is prepared.

In Japan, monkfish liver, or *ankimo*, is a major winter delicacy. According to the classical recipe, cleaning the fresh liver involves placing it in a 3 percent salt brine, to which sake has been added, for about half an hour. Then the large veins are removed. One or more livers, placed in cheesecloth or aluminum foil, are rolled tightly using a bamboo rolling mat to form a thick sausage, which is then steamed for 30–40 minutes. It must be cooled before serving and has the best consistency after being left in the refrigerator for about a day. *Ankimo* is served with a dressing made from soy sauce, *yuzu* juice (or *ponzu* sauce in place of both), dashi, and possibly a bit of sake.

MONKFISH LIVER AU GRATIN WITH CRABMEAT AND VEGETABLES

It is best to presoak the liver under refrigeration for 12–30 hours to remove some of its bitter taste. This can be done in a 3 percent salt brine or, even better, in milk, which seems to be more effective. Next, the liver needs to be tidied up by removing the blood vessels.

Serves 4

250 g (½ lb) monkfish liver, presoaked in milk for 12 hours
500 g (about 1 lb) vegetables, as follows:
　100 g (3¼ oz) ripe tomatoes
　100 g (3¼ oz) porcini mushrooms
　100 g (3¼ oz) red bell peppers
　100 g (3¼ oz) eggplant
　100 g (3¼ oz) celery
1 large clove garlic, puréed
2 shallots
¼ tsp fennel seeds
olive oil
1 Tbsp puréed tomato
1 small chile pepper
150 g (5¼ oz) crabmeat
2 egg yolks
1 small sprig fresh dill, chopped
salt and freshly ground black pepper
8 small pieces grilled bread or brioche
salsa verde from basil, sage, parsley, anchovies, capers, and olive oil
baked tomatoes and dried olives, for garnish

1. Finely dice the monkfish liver together with the vegetables, puréed garlic, chile pepper, shallots, fennel seeds, puréed tomato, and chile pepper to make a coarse *brunoise*. Heat some olive oil in a large skillet over medium heat and cook the mixture, stirring constantly, until it has a uniform consistency.
2. Remove from the heat, allow to cool, and put through a food processor to make a coarse mince. Fold in the crabmeat, egg yolks, and dill. Season well with salt and black pepper.
3. Make a *salsa verde* from basil, sage, parsley, anchovies, capers, and olive oil according to taste.
4. Spread the liver mixture on grilled toast or pieces of brioche and grill in the oven for a couple of minutes.
5. Serve on a bed of *salsa verde* and garnish with baked tomatoes and dried olives.

Tip: Count on using a little less than 1 kg (2.2 lb) of crab claws, of which about one-fifth is actual crabmeat.

▸ Monkfish liver au gratin with crabmeat and vegetables.

FOOD PAIRING HYPOTHESIS

According to this hypothesis, certain combinations of ingredients that share common flavor compounds taste better than those with no common compounds. This has led to surprising examples of pairings, for example, chocolate with caviar, which both contain trimethylamine; and chocolate with blue cheese, which have at least seventy-three flavor components in common. The hypothesis, which is not founded in any scientific rationale, has recently been investigated by a network analysis of 56,498 recipes covering several cultural regions. The investigation showed that, whereas the hypothesis found some support in Western cuisines, the opposite was true for Asian cuisines. The latter seem to avoid combining ingredients that have compounds in common. Hence, food pairing is probably more a matter of tradition and regional food culture than of physiological origin.

▸ Hard cheeses being aged the traditional way, an excellent example of how to make the most of umami.

FOOD PAIRING AND UMAMI

Food pairing refers to a hypothesis that is popular among some chefs. In contrast to the traditional food pairing hypothesis, the principles underlying the synergy in the umami taste, enhanced by teaming up ingredients with glutamate and 5′-ribonucleotides, respectively, is based on a scientific rationale, involving the functioning of the umami receptor. Pairing umami compounds, therefore, has a physiological basis and seems not to be strongly dependent on tradition and regional food culture.

CREATING TASTES SYNTHETICALLY

Building on discoveries about the nature of basic tastes and especially what brings out umami and synergizes with it, researchers have tried to mix together pure substances to mimic the taste of a particular food, for example, shellfish. Opinions vary as to whether or not this has been successful or if it can even be done. After all, there are still researchers who think that it is not feasible to evoke all other possible taste sensations with only the five basic tastes.

Experiments have been carried out to try to reproduce the taste of scallops. Tasters found that, in a synthetic mixture of glutamate, adenylate, glycine, aniline, and arginine, all the components played a part in simulating the taste of scallops with regard to the five basic tastes, as well as their aftertaste and palatability. On the other hand, inosinate had no effect on such a mixture, which makes sense seeing as scallops themselves have none. (See the tables at the back of the book.)

The food industry is very interested in finding as-yet-unknown natural substances or new synthetic substances that can not only improve the nutritional value and potential health benefits of a foodstuff but also impart a more interesting and appetizing taste. Artificial sweeteners are a case in point, as they produce a sweet taste without adding any calories. This is equally true for substances that result in an umami taste or interact synergistically to enhance it. Different chemical modifications of the known 5′-ribonucleotides have been shown to increase the synergistic effect by up to thirty times as much as the effect produced by guanylate.

Molecular investigations of taste receptors are, to a great extent, driven by commercial interests in systematically developing new taste enhancers and food additives. The degree to which many of the already discovered substances can actually activate the umami receptors continues, however, to be a subject of controversy.

Umami: Either as little or as much as you like

A representative selection of raw ingredients and processed food products that contain umami taste substances, ranging from very small quantities (on the left) to an abundance (on the right). The products on the top have basal umami (from glutamate, MSG), while those on the bottom have synergistic umami (from the nucleotides IMP, GMP, and AMP).

Note that the horizontal axes are not linear and the position of a given product on the axis does not correspond to its absolute content of umami substances. Instead, the individual products on each axis are placed in the correct relationship to each other. (See also the tables at the back of the book.)

Basal umami. Glutamate.

Synergistic umami. Nucleotides (IMP+GMP+AMP).

Basal umami
1: cow's milk - 2: apple - 3: carrots - 4: egg
5: pork - 6: Worcestershire sauce - 7: mackerel
8: chicken - 9: green asparagus - 10: caviar
11: green peas - 12: oysters - 13: potatoes
14: ketchup - 15: air-dried ham - 16: miso paste
17: sun-dried tomatoes - 18: walnuts 19: soy sauce
20: dried shiitake mushrooms - 21: anchovies in brine
22: blue cheese - 23: Parmesan cheese - 24: fish sauce
25: Marmite - 26: dried seaweeds (konbu).

Synergistic umami
1: green asparagus - 2: oyster mushrooms
3: sun-ripened tomatoes - 4: crab - 5: beef
6: lobster - 7: dried shiitake mushrooms
8: scallop - 9: shrimp - 10: pork - 11: chicken
12: mackerel - 13: anchovy paste - 14: *katsuobushi*

1 + 1 = 8: Gustatory synergy

*Had we
nothing sweeter
than carrots or milk,
our idea of the quality 'sweet'
would be just as indistinct as is the
case with this peculiar quality [umami].
Just as honey and sugar gave us so clear
a notion of what sweet is, the salts of
glutamic acid are destined to
give us an equally definite
idea of this peculiar
taste quality.*

もし人参あるいは牛乳より甘いものがないなら
ば「甘い」という味の観念はこの独特の（うま
み）という観念の場合と同様に、明確に知ること
ができないであろう。蜂蜜や砂糖が甘味とは何
であるかを教えてくれるようにグルタミン酸塩は
その独特の呈味性(うま味)についてはっきりした
認識を与えてくれる。

Kikunae Ikeda (1864–1936)

Umami from the oceans: Seaweeds, fish, and shellfish

It is likely that many cuisines in all parts of the world originally depended on fermented fish and shellfish, cooked and cured meat, and seaweeds to add umami to a variety of dishes. In both Asia and Europe, preserved fish, together with the condiments made from them, have been used for at least two and a half millennia, and probably since long before then, as a simple, nutritious way to improve the taste of other foods. One might say that the history of using ingredients to prepare food that is rich in umami runs parallel to and reflects the overall evolution of the culinary arts. The heart of the matter is handling the ingredients in such a way that the proteins and nucleic acids are converted to free amino acids and free nucleotides by the skilful use of cooking, brewing, enzymatic fermentation, salting, drying, smoking, and curing, alone or in combination. In the past, these methods were also of great importance to prevent spoilage of the foodstuffs that come from the sea.

SEAWEEDS AND KONBU: *THE* MOTHER LODE OF UMAMI

Before we come to fish and shellfish, let us start our exploration of marine sources of umami with the seaweed kelp, the raw ingredient that contains more free glutamate than any other, up to about three percent, and that was the key to Professor Ikeda's identification of the fifth taste.

The spread of Buddhism from China and Korea to Japan in the sixth century brought with it a very specialized vegetarian cuisine, *shōjin ryōri* or 'devotion cuisine,' about which we will learn more later. It is thought that the idea of using kelp, the large brown alga known as konbu in Japan, to make and impart umami can be traced back to this religious movement.

With the spread in the 1300s of Zen Buddhism, whose monks practiced an even stricter, more ascetic form of vegetarianism that eschewed all animal products, konbu took on additional prominence. It was used to make *shōjin* dashi, the ultimate vegetarian soup stock described in the previous chapter, which is sometimes served with salted dried tofu. As tofu is rich in protein, this simple temple broth is both palatable and nutritious.

Harvesting konbu along the coast of Hokkaido in Japan.

The historical record concerning the commerce in seaweeds gives an indication of the way in which these culinary practices were evolving. Descriptions of the harvesting of wild kelp growing along the coasts of the northern Japanese island of Hokkaido can be found in sources going back to the eighth century. It would appear that the inhabitants of Hokkaido did not themselves have any tradition of using the seaweeds for making dashi, but the monks in the many temples in Kyoto had started to do so. By the fourteenth century, trading ships were carrying sun-dried seaweeds from Hokkaido to Osaka and Kyoto via a 1,200-kilometer, mostly maritime route known as the konbu road.

A WORLD OF KONBU IN JAPAN

Given that konbu is *the* mother lode of umami in Japan, it is hardly surprising that there is an entire science, with deep historical roots, linked to the selection and handling of different types and qualities of the algae. More than 95 percent of all konbu harvested in Japan grows in the cold water along the northern coastlines around the island of Hokkaido. Here alone, more than forty varieties are collected.

Umami from the oceans: Seaweeds, fish, and shellfish

Of the many available types, *ma-konbu* and *Rishiri-konbu* are considered to be the best choices for making dashi because the resulting stocks have a very light color and a complex taste. *Ma-konbu* is the konbu with the largest amount of free glutamate, 3,200 mg per 100 g, whereas *Rausu-konbu* has 2,200 mg per 100 g, and *Rishiri-konbu* has 2,000 mg per 100 g. The lower-quality *Hidaka-konbu* has 1,300 mg per 100 g. And, of course, if they have been aged for a few years, they are even more highly prized.

Two specialty products, *tororo-konbu* and *oboro-konbu*, are evidence of the serendipitous relationship between the city of Osaka and nearby Sakai, an important port on the konbu road. The former is well known for its use of konbu in its distinctive cuisine and Sakai has for centuries been famous for its exceptionally high-quality knives. *Tororo-konbu* is now generally prepared in factories. Dried konbu blades are first marinated for a short time in a rice vinegar mixture, allowed to dehydrate partially, compressed into bundles, and then cut across the fibers into paper-thin shavings by machines with razor-sharp knives. The shavings take on a matte white appearance, tinged with the palest green, and can be as thin as 0.01 mm, less than the diameter of an average human hair. *Oboro-konbu* is made by a similar, but exclusively artisanal, process. After the bundles of marinated seaweeds have been left for a day to become softer, skilled craftsmen scrape them by hand using a knife with a special blade, resulting in shavings that, amazingly, are only slightly thicker than those made by machine. Both types of shavings have a very unusual mouthfeel, not unlike the sensation one gets from eating cotton candy. They are among the softest foods on the planet, so light that they almost melt on the tongue, leaving an aftertaste of umami with a slightly acidic undertone from the rice vinegar. *Tororo-konbu* is often used to enhance the taste of a soup or a tofu dish, while *oboro-konbu*, being cut along the fibers of the seaweed, can be wrapped around cooked rice.

Tororo-konbu (*top*) and *oboro-konbu* (*bottom*).

Like many culinary innovations, *oboro-konbu* and *tororo-konbu* may have been invented by chance. Although dried seaweeds can be preserved quite successfully, they are susceptible to mold if they accidentally become moist under transport or have not been dried sufficiently in the first place. It is thought that people came up with the idea of scraping off the resulting whitish layer and then marinating the seaweed in a little rice vinegar to enhance the taste, kill the mold, and preserve the konbu even longer.

A bundle of dried konbu.

There are many different varieties of konbu, all belonging to the alga order Laminariales. By far the greatest proportion of the farmed konbu is made up of the species *Saccharina japonica*, which is related to the sugar kelp that is common along Atlantic coastlines. The blades of these seaweeds can grow to several meters in length and 10–30 centimeters in width. Most of the konbu sold on the world market is cultivated on ropes in the seas around China and Japan. After harvest, the algae are sun dried, and those of highest quality are put aside to undergo a carefully controlled process of aging in cellars (*kuragakoi*). While the algae can be cured for up to ten years, they are kept, on average, for two years. Curing the seaweeds, in a sense, ripens them so that they lose some of the flavor of the sea and develop a milder taste, allowing their inherent umami and subtler aromas to come to the forefront. In Japan, the best vintage konbu is the object of unabashed admiration, much in the way that wine connoisseurs appreciate the finest bottles from exceptional years.

Konbu contains large quantities of free amino acids, of which 80–90 percent is glutamic acid, but it lacks inosinate and guanylate, which interact synergistically with glutamate. The balance of the amino acid content is made up mostly of alanine and proline, which both impart a sweetish taste to the seaweed. Another substance that contributes to a rich umami taste is sometimes present in the form of a layer of whitish powder that is often exuded onto the surfaces of the blades when they are dried and cured. Under no circumstances should this be washed off, as it is made up of a combination of sea salt, glutamate, and mannitol. Mannitol, which is very abundant in sugar kelp, is what is known as a sugar alcohol and also has a sweetish taste.

Although the various types of konbu are the seaweeds with the most free glutamate, the red algae dulse (*Palmaria palmata*) and laver (*Porphyra* spp.) also contain reasonable quantities of glutamate. Laver, which is used to make nori, also has inosinate and guanylate, with the result that the nori itself is a source of both basal and synergistic umami in sushi.

What is special about konbu and quite a number of other seaweeds is that once they have been dehydrated, and sometimes also cured, the free glutamate is extracted very easily by soaking them in warm water, resulting in umami. Nevertheless, as a good dashi has only about 30 mg of glutamate per 100 g, this process draws out only a small proportion of the total available. With the exception of ripe tomatoes, konbu is probably the raw ingredient that yields the most glutamate with the least preparation.

FRESH FISH AND SHELLFISH

Before we move on to the fermented fish products, we will have a quick look at some foods prepared from fresh fish. Fish and shellfish are unusually good sources both of basal umami derived from free glutamate and synergistic 5'-ribonucleotides. (See the tables at the back of the book.) At the top of the list of free glutamate, we find sardines, squid, scallops, sea urchins, oysters, and blue mussels. Shrimps and mackerel, which contain more glutamate than a variety of types of fish roe, are also near the top. Bony fish, especially those with darker meat, have significantly less free glutamate.

Anchovies, sardines, scallops, squid, mackerel, tuna, and shrimps are high on the list of foods with an abundance of nucleotides, and in particular, inosinate and adenylate. But as we will see later, there is a much lower concentration of the substances that impart umami in fresh fish than in their dried and fermented counterparts. Smoked fish normally also have a stronger umami taste than fresh fish.

Just because a fish or a shellfish is rich in umami substances, it does not follow that this will be the most prominent taste. For example, sea urchins contain the amino acid methionine, which has a bitter, characteristic sulfurous taste that is redolent of the sea. When that is taken away, sea urchins taste more like crab or shrimps. Another example is crab, which has a great deal of the amino acid arginine, which imparts a taste that is both bitter and salty like seawater.

COOKED FISH AND SHELLFISH DISHES AND SOUPS

Gentle cooking and steaming of fresh fish and shellfish produce umami by releasing their free glutamate and, more important, the nucleotides that interact synergistically with it. ⋯> *Pearled spelt, beets, and lobster (page 70)*. This is why it is quite possible to make delicious soups using only these two ingredients. Even more umami can be achieved by adding vegetables and herbs to a simple shellfish bisque. ⋯> *Crab soup (page 76)*. A savory combination of fish, shellfish, vegetables, and seaweeds is to be found in the traditional clambake made on the beach and in a modified version prepared in a pot. ⋯> *Clambake in a pot (page 78)*. The traditional French bouillabaisse is a thick soup made with fresh fish, shellfish, and often vegetables and eggs. It is served with a dollop of *rouille*, a sauce that is thickened with an egg yolk. It is a felicitous combination of ingredients that are filled with umami substances.

PEARLED SPELT, BEETS, AND LOBSTER

Serves 4

2 live lobsters weighing about 300 g (⅔ lb) each
olive oil
assortment of coarsely chopped vegetables, such as leeks, carrots, celery, parsley stalks
tomato paste
a little dry white wine
2–3 Tbsp butter

200 g (⅞ c) pearled spelt
8 small beets
1 shallot, finely chopped
pinch of cayenne pepper
3 dL (1¼ c) chicken bouillon
60 g (1⅔ oz) finely grated Parmesan
salt and freshly ground black pepper
fresh tarragon leaves

1. Pierce the lobsters through the head. Remove the heads and the claws. Split the tails into two pieces, leaving the shells on, and remove the innards.
2. Fry the claws and tail pieces in a pan in warm olive oil until a real shellfish aroma develops. The lobsters should only be browned—the meat should not be cooked through. Remove the meat from the claws and the tails.
3. Cut the heads into chunks, toast them in a skillet in a little warm olive oil together with the shells from the tails and claws. Add the assorted vegetables, tomato paste, and wine. Reduce the liquid.
4. Add water just to cover the shells, bring to a boil, and skim off the foam.
5. Add 2 or so tablespoonfuls of the butter, cover, and allow to simmer on low heat for about half an hour. Pass through a fine-mesh sieve and refrigerate.
6. The butter will solidify and rise to the surface. Remove a little of it and put in a pot. Reserve the rest of this lobster butter for sautéing the lobster meat. Keep the lobster bouillon for cooking the pearled spelt.
7. Rinse the spelt. Peel the beets very carefully, and if necessary, cut them into smaller pieces.
8. Warm the lobster butter in the pot, add the finely chopped shallot, and fry gently without allowing the shallot to brown. Add the drained spelt and cayenne. Add the chicken bouillon and lobster bouillon alternately, a little at a time. Place the beets in the spelt.
9. Cook the spelt, pricking the beets from time to time to test for doneness. Remove the beets when they are soft but not overdone.
10. Allow the spelt to continue simmering. It is ready when it is still a little firm. Mix in the Parmesan cheese, season with salt and pepper, and return the beets to the pot. Add the tarragon.
11. *To serve:* Sauté the lobster meat in the remaining lobster butter in a skillet over medium heat just before serving. Arrange on plates with the spelt and beets. Drizzle with any lobster butter remaining in the sauté pan.

▸ Pearled spelt, beets, and lobster.

Umami and the art of killing a fish

Ikijime, which means 'to terminate while alive,' is a 350-year-old Japanese technique for killing fish. It has the effect of delaying the onset of *rigor mortis*, thereby ensuring that the taste of the fish is of the highest quality and that there is least damage to, and discoloration of, the flesh. The fish dies humanely and unstressed, which preserves and releases more of the savory substances that bring out umami.

The traditional method is as follows. With a heavy knife, a cut is made in the head on the dorsal side of the live fish, slightly above and behind the eyes, severing the main artery and the elongated medulla, which is the lowest part of the brain stem. This is the part of the brain that controls movement. A second cut is made where the tail is attached to the body. Then the fish is plunged into an ice slurry in order to allow it to bleed out. The muscles of the fish relax in the ice-cold water while the heart continues to pump, but the fish has ceased to struggle for its life and is unstressed.

The final, definitive step is to shut down completely the autonomic nervous system, which continues to send messages to the muscles to contract. It is destroyed by inserting a long, very thin metal spike along the length of the fish through the neural canal of the spinal column. At this point, the fish relaxes totally and all movement ceases.

The blood that remains in the muscles retracts into the entrails of the fish, which are removed under running water so that blood and digestive fluids do not spill onto the flesh. The head, tail, gills, and fins are cut off and the fish is wrapped in paper or cloths to absorb any blood that might still seep out. At this point, the fish can be filleted for cooking, sliced for sashimi, or allowed to age for one or two days in the refrigerator.

Surprisingly, a really fresh fish is not always the one that tastes best. Allowing the fish to age generally brings out a greater range of taste impressions and more umami because taste substances are released into the muscles. At the same time, the ongoing enzymatic breakdown of the muscle fibers at low temperatures leads to a softer, more pleasing texture and a much better mouthfeel. In the case of flatfish, for example, this is partly due to the release of inosinate, which interacts synergistically with the glutamate content. Naturally, the determining factor is whether the fish has been killed by *ikijime* so that the fillet is perfect, with no traces of blood or digestive fluids. There should also be no signs of tissue damage caused by the trauma or rough handling that are characteristic of less skillful ways of slaughtering and bleeding out the fish. Fish that have been killed by *ikijime* and then allowed to age are known in Japanese by the term *nojime*.

At first sight it might appear that *ikijime* is a brutal technique, but there is no doubt that if it is carried out professionally it is a very humane way of killing a fish and causes it the least suffering. At the same time, it allows the fish to be used to the best advantage, with more taste and higher gastronomic value.

▸ Chef Toshio Suzuki demonstrates *ikijime*, the traditional Japanese technique for killing a fish, at the Gohan Society in New York.

A traditional clambake: New England method, Danish ingredients

What makes a traditional New England clambake so special? Of course, a festive occasion and the fresh sea air are part of the equation, but there is more to it. Clambake is a way of cooking fish, shellfish, and vegetables between layers of fresh seaweeds in a stone-lined hollow on the beach. This is a wonderful example of a recipe that has evolved into a simple way to maximize the intensity of the fifth taste. This is all due to the seaweeds. As they warm up, they release glutamate, which interacts synergistically with the ample quantities of nucleotides found in the fish and shellfish.

The appeal of trying to re-create a typical New England clambake using local Danish ingredients was not lost on a small group of our seafood-loving friends who joined us to do so. On a summer day in August, we headed for a quiet beach on the northern part of the island of Funen. Two days beforehand, some of us had slaved away to construct a stone oven in the sand. It consisted of an oval hollow, about 80 centimeters deep and 1 meter across at the widest part. The bottom and sides were lined with large stones, placed as close to each other as possible.

To start the process, a fire was lit in the hollow and for the next three hours we kept adding more wood to heat all the stones, both on the bottom and at the sides. In the meantime, we collected pieces of fresh bladder wrack, enough to fill the entire hollow. When the wood had almost burned completely and the stones were red-hot, we removed the largest of the embers and remaining bits of charred wood. The oven was now ready to be filled with all of the good things that we had on hand: mussels, lobsters, langoustines, a large catfish, pike-perch, ocean perch, mackerel, corn on the cob, and potatoes. As you might have noticed, the vegetables were chosen with an eye to their glutamate potential. The only thing missing was clams!

First we placed a thick layer of bladder wrack in the bottom and then added our ingredients in layers, with more seaweeds in between them. Those ingredients that needed the longest cooking time were placed at the bottom. A final layer of seaweeds was added to form a sort of flat lid, on which half a dozen fresh eggs were placed. They act like a thermometer to indicate when the clambake is ready to eat. More wet seaweeds were placed on the stones that formed the rim of the oven in order to prevent the canvas tarpaulin, which was placed on top to keep the steam in, from scorching or catching fire. Once everything was in place, what was required was patience. In the meantime we made a bisque from small live crabs that we found on the beach. ▸ *Crab soup (page 76)*. After about an hour and a half, we took one of the eggs out from under the canvas. It was not yet cooked through, which meant that the clambake was also not ready, and we had to resign ourselves to waiting a little longer. Half an hour later, we pulled out another egg. It was exactly the consistency of a soft-boiled egg, a sign that we could start to unpack the oven.

Once we had everything on the table and had poured glasses of dry white wine, we tucked in. The food tasted absolutely wonderful—it was as if the umami was rising out of it like a vapor. The clambake was a total success and turned out to be well worth both the effort and the long wait.

▸ Clambake at a Danish beach

CRAB SOUP

Serves 4

- 2 kg (4½ lb) small live crabs
- 100 g (3½ oz) celeriac
- 100 g (3½ oz) onion
- 100 g (3½ oz) leek tops
- 100 g (3½ oz) carrots
- ½ dL (⅕ c) olive oil
- 3 dL (1¼ c) dry white wine
- 1 dL (⅖ c) dry white vermouth, such as Noilly Prat
- 5 Tbsp tomato purée
- 2 bay leaves
- 1 sprig fresh thyme
- 1 head fresh dill
- ½ chile pepper, chopped
- 1 tsp light brown sugar
- 1 L (4¼ c) water
- 100 g (3½ oz) butter
- salt and freshly ground black pepper
- possibly a little Tabasco sauce
- bread for making croutons
- olive oil
- 1 large clove garlic, crushed

1. Sort the crabs carefully and discard dead ones, if any. Even one dead crab can spoil the soup. Cut the vegetables coarsely into cubes.
2. Lightly sauté the crabs in very hot oil in a pot with a thick bottom.
3. Add the vegetables and stir to coat with the oil, then add the wine and vermouth, together with the puréed tomato, bay leaves, thyme, dill, chile pepper, and brown sugar.
4. Cook until the liquid has reduced by two-thirds, then add the water and simmer for 30 minutes.
5. Strain off the bouillon, crush the crabs in the pot with a heavy spoon, and return the stock to the pot. Allow the soup to simmer for another 30 minutes, adding the butter halfway through.
6. Carefully skim off the butter and reserve to add color and taste to the soup at the end. Strain again to produce a clear bouillon to use as a base.
7. Season with salt, pepper, and possibly a little Tabasco sauce, and whisk in a little of the crab butter.
8. To make croutons, trim the crusts off the bread, cut into small cubes, and toast them in a pan with hot olive oil and the crushed garlic clove. Serve the soup garnished with the croutons.

THE FULL-BODIED VERSION

Mix together saltwater fish and langoustines cut into chunks, mussels, a little tomato paste, olive oil, parsley, garlic, finely chopped onion, and salt. Cook in the soup for 1–2 minutes, so that the fish is still a little raw in the middle. Served steaming hot and topped with bread, garlic mayonnaise, and a head of fresh dill, this makes a main course.

▸ Small crabs in the soup pot.

Clambake in a pot.

CLAMBAKE IN A POT

If you do not have access to a beach, or one with stones, or one where you are allowed to build a fire, you can still make a very successful clambake. You can cook juicy lobsters, clams, or a whole fish on a bed of seaweed in a huge pot on the top of the stove, where you can control the temperature. Here we give instructions for a version with lobsters.

First you need to gather fresh, living seaweeds, such as bladder wrack, from an area where you are sure that the water is clear and unpolluted. The seaweeds need to be rinsed thoroughly to remove any sand, small shells, or sand fleas that might be on them. Then place a layer that is 15–20 centimeters (6–8 inches) deep in the bottom of the pot. Place two live lobsters back to back on top of the seaweeds. Insert a thermometer attached to a cord in the middle between the lobsters and cover everything with another layer of seaweeds of the same size.

Cover the pot with a tight-fitting lid so that no steam can escape. Turn the stove element to the lowest possible heat setting and heat to a temperature of 65°C (150°F). This should take about an hour and a half. The lobsters are now cooked, already salted to perfection by the seaweeds. Serve with a small crisp salad, toasted bread, and a homemade mayonnaise to which a bit of Worcestershire sauce has been added. It is a meal fit for a king.

The liquid remaining in the pot can be reduced to make a sauce with strong tastes of lobster and umami. It can be used, for example, to nap the lobster meat or to season the mayonnaise.

Excavation of a Roman *garum* factory in Almuñécar, Spain.

EVERYDAY UMAMI IN ANCIENT GREECE AND ROME

Even though fresh fish and shellfish are good sources of umami, their effectiveness is greatly enhanced by drying or fermenting them. In fact, fermented fish contain more inosinate than any other foodstuff. From time immemorial, people in many parts of the world have been making them into sauces and pastes for use as taste enhancers. In Europe, the earliest known example is a traditional fish sauce that was greatly valued by the ancient Greeks and Romans as a universal condiment. It is indisputably the oldest additive used in the West to impart umami.

The Romans were very partial to this salty, fermented fish sauce, called *garum,* and later also *liquamen,* in Latin. It was used both as a salt seasoning and to add deliciousness—namely, what we know as umami. Classical Roman works on the culinary arts suggest that *garum* could be incorporated into just about any dish, even sweet soufflés. Probably, however, it was most often mixed with something else, for example, with wine vinegar (*oxygarum*), with honey (*meligarum*), with wine (*oenogarum*), or with spices.

The etymology of the word *garum* is a bit of a mystery. It is derived from a fish known to the ancient Greeks as *garos*; Pliny the Elder includes it in his list of the fish found in the oceans, but without any further description. The sauce was made on the Aegean islands at least as early as the fifth century BCE, and eventually its production spread to many of the fishing towns all around the Mediterranean. No sources describe exactly how it was made, only that it was made from fish and that it smelled horrible. While early writers sometimes characterize it as putrid, in other writings they express an enthusiastic fondness for the sauce, referring to its "very exquisite nature."

It seems that the term *garum* applied only to the brownish liquid that seeped out when small fish and fish intestines, especially those from mackerel and tuna, were salted, crushed, and fermented. The most highly prized *garum* was made from the blood and salted innards of fresh mackerels, which were fermented for two months. It is said that the mackerels had to be so fresh that they were still breathing. A lower quality sauce, *muria*, was made from the briny liquid drawn out when tuna was preserved in salt.

Fermentation took place outside in the dry heat. The stench must have been awful, and that is probably why no ruins of *garum* factories have been found inside the walls of Pompeii. There were, in fact, regulations forbidding the erection of establishments for the production of *garum* within a distance of three stades (or a little more than half a kilometer) from a town. Nevertheless, *garum* production was so widespread and such a valuable source of income that the welfare of many coastal towns depended on it. The sauce acquired a status that put it on a par with olive oil and wine. It was stored in tall, slender pitchers or in jars, which can probably best be compared to contemporary bottles with ketchup or Worcestershire sauce. *Garum* may have been mixed with olive oil at the dinner table to make what we would now think of as a vinaigrette dressing.

Some of the Roman writers thought, erroneously, that *garum* is the liquid that is formed when fish rots, due to bacterial decomposition. Actually, it is the result of fermentation, in which salt and the enzymes of the fish itself cause it to break down. In this, the innards of the fish are very important, as the intestines contain large quantities of proteolytic enzymes, which work on the proteins in the fish. This process is accelerated by using salt, which draws out the liquid from the fish and, furthermore, inhibits the growth of bacteria.

The fermentation process undoubtedly released great quantities of free amino acids, among them glutamic acid. It involved some combination of the flesh, innards, and blood of fish, but other than that many of the details of how *garum* was made in antiquity have never been discovered. It seems that there were local variations as to which types of fish and what parts of them were used and in what proportions. The sauce could contain large amounts of fish, especially if it was made from small anchovies, which almost liquefy under fermentation. It is possible that *garum* was made with blood and *liquamen* with whole fish. Even though the paste-like dregs, called *allec*, that were left behind when the liquid was skimmed off were malodorous, the *garum* itself appears to have had a mild, pleasant taste.

The *garum* trade reached its peak from the second to the fourth centuries CE, but *garum* never lost its appeal in the Mediterranean region, where it was used throughout the Middle Ages and Renaissance. It has survived to this day in the form of salted anchovies and anchovy paste, both of which have a cleaner taste because the innards are removed from the fish before they are processed. The different varieties of fish sauces now made in Southeast Asia—for example, oyster sauce—all contain great quantities of glutamate and can be seen as counterparts to the Roman *garum*, just as Worcestershire sauce can be considered to be its modern Western offshoot.

The use of *garum* has been immortalized in the legendary work concerning the culinary arts, *De re coquinaria* (*On the Subject of Cooking*), the oldest book of recipes that has survived from European antiquity. It is popularly attributed to a Roman gourmet and lover of luxurious things, Marcus Gavius Apicius, who lived in the first century CE. As the text was probably compiled sometime around 400 CE, it is doubtful that the work is actually by a true historical figure. More than 80 percent of the approximately 500 recipes in the book incorporate *garum*, in many cases as a substitute for salt. ⇢ *Patina de pisciculis (page 82)*

FISH SAUCES AND FISH PASTES
Even though we have documentary evidence that the ancient Greeks and Romans used *garum* as a fish sauce more than 2,500 years ago, we probably have to look to the cultures of the East, and especially to that of China, to find the origins of contemporary fish sauces. It was mostly the Chinese living in the coastal areas who made fish sauce, probably by combining small fish with the innards from larger cooking fish, causing fermentation. Once soybeans were added to the salted and fermented fish in order to stretch the ingredients further, the liquid started to evolve into what would later become soy sauce, which came to be used much more widely, particularly inland.

Fermented sauces and pastes made from fish, shellfish, and mollusks—possibly more than 300 different types, covering a broad spectrum of qualities and price levels—are common throughout most of Southeast Asia. They are known by a number of different names: *yu-lu* (China), *nuoc mam tom cha* (Vietnam), *nam-pla* (Thailand), *teuk trei* (Cambodia), *nam-pa* (Laos), *patis* (Philippines), *bakasang* (Indonesia), *ngan-pya-ye* (Myanmar), *budu* (Malaysia), and *ishiri* (Japan). Almost all fish sauces are made according to the same basic method. It involves the salting and fermenting of either whole fish and shellfish or parts thereof, including blood and innards. Fermentation takes place using the enzymes found in the animal products, releasing an abundance of free amino acids, especially alanine and glutamate. Some are made with fresh fish, others with ones that have been dried first; anchovies are included in most of the sauces. As an example, *ishiri* is made from fermented sardines, other small fish, and the innards from squid. When the sauces are fermented for a long period of time, they exhibit tastes similar to those of nuts and cheese. Modern Thai fish sauce is fermented for up to 18 months, much longer than the Roman *garum* of antiquity, and it also has a considerably higher salt content. Fish sauces have come to be associated with some very traditional dishes. In Korea, *shiokara*, made from fermented salted fish, is added to that country's best known dish, kimchi, which is fermented cabbage sometimes mixed with other vegetables. Similarly, in Japan, *ika no shiokara* consists of small squid fermented using the enzymes from their own innards.

PATINA DE PISCICULIS

This recipe for a fish dish is based on the ancient Roman one by Apicius. The original version does not specify which type of fish was used. Here we have chosen small sprats, but you could also use small pieces of salmon, or small herring, sardines, or anchovies. It is actually unlikely that herring or salmon would have been available in Rome. Also, the original recipe called for twice as much oil as this version.

Serves 4

400 g (a little less than 1 lb) small fresh sprats (about 20 fish)
salt
½ tsp freshly ground black pepper
all-purpose flour, for dredging
1 dL (⅖ c) olive oil

2 onions, chopped
150 g (5¼ oz) raisins
2 sprigs fresh lovage
1 Tbsp fresh oregano leaves
½ dL (⅕ c) garum or fish sauce

1. Season the fish with the salt and the pepper and coat with the flour.
2. Panfry the fish in the oil in a large skillet over medium heat. Remove and keep them warm.
3. Add the onions to the oil and cook until they are translucent.
4. Add the raisins, lovage, oregano, and *garum* or fish sauce to the onions and season to taste.
5. Pour the oil mixture over the warm fish just before serving.

▸ *Patina de pisciculis.* Classical Roman fish dish with *garum*, based on the ancient culinary work *On the Subject of Cooking*. On the table are flasks with *garum*, olive oil, and wine, a feature of every meal in ancient Rome.

The most important contributor to umami in fish sauces is glutamate. The content of free glutamate can be very great, close to 1,400 mg per 100 g in Japanese *ishiri* and the Vietnamese *nuoc mam tom cha*, or a little more than in soy sauce. Usually, the salt content is also very high, all the way up to 25 percent, which is quite a bit more than the 14–18 percent found in soy sauce. The combination of umami and salt works synergistically to enhance the saltiness of fish sauce.

Interestingly, in contrast to such other taste additives as anchovy paste and certain original Asian fish extracts, fish sauce has no significant content of free 5'-ribonucleotides. In ancient Japan, a concentrated fish extract, *irori* or *katsuo-irori*, was produced by reducing to a paste the liquid in which bonito (*katsuo*) had been cooked, using the same techniques as those now used to make *katsuobushi*. *Katsuobushi* is the component of classical Japanese dashi that contributes the 5'-ribonucleotides that interact synergistically with the glutamate from konbu to impart a strong umami taste. As *katsuobushi* dates back only to the 1600s, it is a much newer arrival on the culinary scene than *irori*.

Although *irori* disappeared from Japanese cuisine during the Meiji era (1867–1912) and is no longer on the market, a similar product, *senji*, is still made on the island of Kyushu in the southern part of Japan. When produced according to the traditional recipe, *senji* contains 900 mg of glutamate per 100 g and no less than 786 mg of inosinate per 100 g. Comparably high levels of inosinate are found in another traditional fish extract, *rikakuru*, from the Maldives. Both of these sauces are much richer in inosinate than *katsuobushi*, which has only about 474 mg per 100 g.

In Asia, fish sauces are generally incorporated into cooked dishes or served as a condiment with, for example, rice. They are much more widely used as salt substitutes and taste enhancers than as true sauces to be poured over food when it is served. The fish sauces can be seasoned with other ingredients, such as chile and lime juice.

The quality, and with it the content of substances that impart umami, of all of these fish sauces and pastes is very dependent on the raw ingredients that go into them and the method of production. Many of the commercially available products contain a great deal of salt and added MSG, and they do not live up to the high standard that is the hallmark of traditionally made fish sauces.

Commercially produced Asian fish sauce.

Cooking fresh mackerel in salt for production of quick-and-easy *garum*.

MODERN *GARUM*

There are a number of recipes for the classical Roman fish sauce *garum*. They are quite simple and can easily be made with raw ingredients that are widely available, but an authentic *garum* needs to ferment for several months in heat and sun, which can be a problem in some climates. Let us not even begin to talk about the stench! Mackerels are a good choice, because these fish contain very aggressive enzymes in their innards, leading to powerful fermentation and the production of a great deal of glutamate. *Garum* can also be made from just the fish entrails. Experiments using just mackerel innards have resulted in a sauce with 623 mg glutanate per 100 g. ⇢ *Garum (page 86)*

As you can see, making a true *garum* is a slow process. But there is a shortcut, which survived from a Greek source about agriculture (*Geoponica*) compiled in the tenth century CE from even older works. In it we find a recipe for a sort of quick-and-easy *garum* for the benefit of "the busy Roman housewife." This recipe calls for cooking salted fish and innards for about two hours. As it is made without any fermentation, it is milder than traditional *garum* and imparts much less umami.

GARUM

2 kg (4½ lb) very fresh mackerels   *400 g (1¾ c) salt*

Cut the whole mackerels, innards and all, into smaller pieces and place them in a crock. Mix in the salt. Cover the crock with a loose-fitting lid and place it outside where it is warm, preferably in the sun. Allow the fish to ferment for a couple of months, turning the fish pieces in the salt once in a while. Then strain out the liquid, first in a sieve or colander to remove the large pieces and then through cheesecloth to ensure that the liquid is completely clear. Store the finished sauce in the refrigerator in a bottle with a stopper. If you wish, you can dilute it with a little leftover wine. For this recipe, you can also use just the entrails and blood from mackerels.

On account of the significant salt content, which is about 20 percent, both the fermented *garum* and the quick-and-easy *garum* can, in principle, be kept indefinitely in a closed container in the refrigerator. ⋯> *Quick-and-easy garum*

If you want to make a quick-and-easy *garum* with more umami, you can add water from cooking potatoes, tomato juice, and smoked fish or fish skin. This boosts the glutamate content to 217 mg per 100 g. ⋯> *Smoked quick-and-easy garum*

QUICK-AND-EASY *GARUM*

2 kg (4½ lb) very fresh mackerels   *400 g (1¾ c) salt*
1 L (4¼ c) water   *sprigs fresh oregano*

Cut the whole mackerels, innards and all, into smaller pieces and put them in a pot together with the water, salt, and some oregano. Bring the water to a boil, turn the heat to low, and allow the fish to simmer for about 2 hours. Then strain out the liquid, first in a sieve or colander to remove the large pieces and then through cheesecloth to ensure that the liquid is completely clear. Store the finished sauce in the refrigerator in a bottle with a stopper.

Quick-and-easy *garum*.

SMOKED QUICK-AND-EASY GARUM

4 large onions, unpeeled
2 kg (4½ lb) very fresh whole mackerels, cut in chunks with their entrails
1 piece skin from smoked mackerel or salmon
2 L (8½ c) water used to cook mature potatoes
1 L (4¼ c) juice from the pulp of tomatoes
½ L (ca. 2 c) fresh oregano leaves
300 g (1⅓ c) sea salt

1. Cut the onions in half. Scorch them in a hot pot over high heat without any fat or oil until the cut side is completely black.
2. Add the mackerel, fish skin, water, tomato juice, oregano, and salt, and simmer on low heat for about 2 hours.
3. Strain through a fine-mesh sieve or cheesecloth. Return the liquid to a (clean) pot and cook to reduce the volume to 1 L (4¼ c).
4. Pour the *garum* into a bottle with a stopper. Store it in the refrigerator.

SHELLFISH PASTE

Fermented shellfish sauces and pastes are made all over Southeast Asia, often from shrimps and oysters. Shrimp sauce and paste are used in the preparation of most meals, especially fish and vegetable dishes, in such countries as Thailand, Myanmar, China, and the Philippines. Both are made from tiny shrimps that are washed, mixed with salt, crushed, and then dried in the sun. After a while, the fermentation process converts the shrimps first to a runny liquid and then into a thicker dark purple or brownish paste with a pungent smell. In Myanmar, this is called *ngapi*. Because it needs no refrigeration, it is often sold in the shops or markets in solid blocks. *Ngapi* is a fantastic source of umami, as it has 1,647 mg free glutamate per 100 g.

OYSTER SAUCE

Oyster sauce deserves a special mention, even though the most commonly available commercial products have a poor reputation because they are oversalted and full of added MSG. A true oyster sauce is made by a slow reduction of water in which oysters have been cooked. The result is a thick, caramelized brownish liquid, to which no salt is added, and consequently there is no fermentation. Nevertheless, oyster sauce has large quantities of free glutamate, typically 900 mg per 100 g, which is due to the large glutamate content of the fresh oysters themselves.

Sadly, many commercial oyster sauces are made with oyster essence in water, thickened with cornstarch. Caramel coloring, salt, and MSG are then added. A special variety for vegetarians is made with mushrooms instead of oysters.

SUSHI AND FERMENTED FISH

It will probably come as a surprise to most sushi lovers that originally sushi was not made with fresh raw fish, but with fish that had been fermented for up to half a year. This was partly out of the necessity, in earlier times, to be able to preserve fish in such a way that it could be stored and transported. In all likelihood, sushi originated in several places in Southeast Asia, but the first known mention of it is in a Chinese dictionary dating from the third century BCE. It is thought that it was introduced to Japan about a millennium later.

This early form of sushi was made by cleaning and salting fresh fish, which were then placed in layers in barrels interspersed with layers of cooked rice. A stone was placed on top of the lid of the barrel to press it down. The fish started to ferment due to their own enzymes and the lactic acid bacteria that thrive in the salty, starchy cooked rice. After a few months, the rice was no longer edible, but the fish had kept their nutritional value, even though their taste was much different from that of fresh fish. This type of sushi is called *nare-zushi*, meaning sushi that has been cured. Over time, the Japanese refined the technique. One of the innovations, around the beginning of the seventeenth century, was to add rice vinegar, which tenderized the fish and allowed the fermentation period to be shortened.

Modern sushi, known as *nama-zushi*, which means 'raw sushi,' is really only symbolically related to its original precursor. The fish is now very fresh and usually raw, and the rice is seasoned with a marinade made from rice vinegar, sugar, and salt in order to take on the traditional taste. *Nama-zushi* has much less umami taste than *nare-zushi*, as it is due solely to the free glutamate and 5'-ribonucleotides found in the raw fish and shellfish that are placed on top of the freshly cooked rice. Sometimes the chef might impart umami to the rice by adding a piece of konbu or some *katsuobushi* to the cooking water. In addition, many types of sushi are prepared with nori, which contains a great deal of glutamate as well as 5'-ribonucleotides. Finally, the pieces of sushi are often dipped in soy sauce, which is really just umami in concentrated form.

▸ *Gunkan-maki* with nori and salmon roe.

FUSHI: FISH PRESERVED FIVE DIFFERENT WAYS

Fushi is a term that denotes fish that are preserved as described for *katsuobushi*. What the different types of fish used to make *fushi*—for example, bonito (*katsuobushi*) and tuna (*maguro-bushi*)—have in common is that they must not be too fatty, because this makes it harder to dry the fish and it can end up with a less delicate taste. Mackerel, for example, which is very oily, is less well suited for making *fushi*. On the other hand, *fushi* made from tuna has a very mild taste and, when combined with high-quality konbu (e.g., *Rishiri-konbu*), yields the finest dashi.

KATSUOBUSHI

The combination of *katsuobushi* and konbu are intimately associated with the concept of umami. *Katsuobushi* is rich in inosinate, about 474 mg per 100 g, which interacts synergistically with the glutamate in konbu. It is produced from bonito, called *katsuo* in Japanese, a fish related to mackerel and tuna. Bonito is preserved using no less than five different methods: it is cooked, dried, salted, smoked, and fermented. The great quantities of 5'-ribonucleotides contained in it are the result of this laborious and meticulous treatment. On the other hand, *katsuobushi* has only a little glutamate, about 23 mg per 100 g.

A forerunner of *katsuobushi* was bonito that has only been dried, and in the eighth century CE this expression simply meant 'solid dried fish.' In 1675, Youchi Tosa figured out how to improve the taste by smoking the fish and allowing a mold to grow on it. Today, *katsuobushi* is manufactured in several coastal towns in Japan, including Yaizu, Tosa, and Makurazaki. In the latter, there are seventy small family businesses that make it from freshly caught bonito. A characteristic smell of smoke and fish hangs over the entire city.

There are two main types of *katsuobushi*. One, *arabushi*, is not fermented and has the milder taste, and the other, *karebushi*, is fermented and much harder. There are also two types of *karebushi*. One type still has its red bloodline on the side of the fillet, and on the other the bloodline is cut off (*chinuki katsuobushi*). The one without the red part has a milder, less bitter, and more delicate taste.

Two types of *katsuobushi* in different forms: whole *arabushi* fillet and *arabushi* powder shown together with Hon-dashi powder (*top*) and whole *karebushi* fillet and *karebushi* shavings (*bottom*).

> ### CATCHING *KATSUO* TO OPTIMIZE UMAMI
>
> In order to make the best tasting *katsuobushi* with a high inosinate content, it is important not to stress the fish. Otherwise, all the ATP (adenosine-5'-triphosphate) in its muscles are depleted by trashing around in a net.
>
> In earlier times, *katsuo* were caught near the Japanese coast by fishers in small boats using poles and lines with a hook. It was a very gentle way, and it tired the fish as little as possible. Now the fish are caught on a large scale with the help of a particularly careful technique. A circular net is spread around and under a school of *katsuo*, the fish are brought on board, and they are immediately deep frozen in saltwater.

Niboshi: small cooked sun-dried fish.

NIBOSHI

Niboshi are another source of the 5'-ribonucleotides (in particular, inosinate) that enhance umami. This product is made from small fish, such as anchovies and sardines, or the young of larger fish, such as flying fish, some species of mackerel, and dorade. In contrast to bonito, these types of small fish are often quite oily, which results in a stronger fish taste than that in *katsuobushi*. Although they can be used to make dashi, the result is less delicate, but quite adequate for daily use or in a miso soup.

To make *niboshi*, the fish are first cooked for a short time in a mixture of salted water and seawater. In some cases the fish are also smoked or grilled and air dried (*yakiboshi*), which results in a stronger taste. The fish are then dried. Traditionally, this was done outdoors in the sun, but now industrial methods are employed. Apart from concentrating the taste of the fish and releasing amino acids and inosinate, this processing technique ensures that *niboshi* have a long shelf life, as long as they are kept dry or frozen.

The hardest foodstuff in the world

The hunt for the hardest foodstuff in the world, katsuobushi, took me, Ole, on a trip to Yaizu, a coastal town in Shizoka Prefecture in Japan, only about an hour from Tokyo on the wonderful Japanese bullet train, Shinkansen. Yaizu calls itself "Japan's fishing city," and this is where I would see how katsuobushi is made. My guide was Dr. Kumiko Ninomiya, one of Japan's most prominent umami researchers, and we were joined by Mio Kuriwaki and Dr. Ana San Gabriel from the Umami Information Center in Tokyo. Ana San Gabriel is also a well-known umami researcher. Among her discoveries are umami receptors in animal stomachs; more recently, she has been studying umami taste in breast milk substitutes.

Dr. Ninomiya is the director of the Umami Information Center. She has contacts among those people in Yaizu who can make it possible for a foreigner like myself to gain admission to the harbor area and to a factory, Yanagiya Honten, where katsuobushi is made. Tooru Tomimatsu, president of the organization Katsuo Gijutsu Kenkyujo, showed us around.

We were lucky to arrive on a day when the frozen raw ingredients, katsuo (bonito), were being unloaded from the large fishing vessels, which had been at sea for a month. They had sailed as far as the southern Pacific Ocean and the seas around Micronesia in order to catch this highly desirable fish.

The rock-hard fish, which are frozen to -30°C, each weigh 1.8–4.5 kilograms. They are sorted right on the quay, at first automatically and then by an army of workers who take over and sort them again carefully by hand. The fish are then transported to the factory for processing.

Step one is to defrost the fish in water that is aerated with air bubbles. The temperature must not rise above 4.4°C in order to prevent the important taste substance inosinate from breaking down. Inosinate is absolutely central to the umami taste of the finished katsuobushi. Once they have thawed, the heads of the fish are cut off and the entrails removed mechanically. These castoffs are ground into a mush and used to make fish sauce. The fish is now simmered at 98°C for almost two hours in brine that is used over and over. This brine accumulates many of the substances that impart the desired taste to the finished product. As the fish simmer, their proteins are denatured, the flesh becomes firmer, and the amount of free inosinate increases approximately thirtyfold. The cooked fish are cut up into fillets, trimmed, deboned, and skinned by hand.

Now comes the most important part of the process: namely, dehydrating the fish and reducing the water content from 65 percent to 20 percent, first by drying and smoking and then, in some cases, by fermentation.

The real secret lies in how the fillets are smoked. Here our timing was also lucky. Just as we arrived at the area where the four-story-high smoking ovens are located, the 'smokemaster' was ready to start fires in the special places called hidoku that lie at the bottom of each oven. We eased ourselves through the little trapdoor in the floor and climbed down a steep ladder just in time to see him light the fires in an array of large circular smoke basins where the firewood had been piled up. By crouching down on the floor of the oven, we were able to es-

Unloading frozen bonito (*katsuo*) for *katsuobushi* production.

experience the tension in the air as the wood burst into flames and the process began. Then it was a question of scrambling quickly up the ladder and out of the oven before the large trapdoor shut.

Only two types of hard oak wood (*konara* and *kunugi*, a type of chestnut oak) are used, as they impart the exact smoky taste that the Japanese prefer in their *katsuobushi*. The firewood is replenished and the fires relit up to four times a day. The company guards its trade secrets very well, and the inside of the oven was the only place where I was not allowed to take photographs.

The fish are arranged on wire trays stacked in the bottom of the smoke oven to dry out for about a day, reducing their water content to 40 percent. Then the trays are moved up to the topmost stories of the oven and smoked for ten days, thus reducing their water content even further, to 20 percent. The finished fillets are *arabushi*, one of the several varieties of *katsuobushi*. This is the product that is referred to as the hardest foodstuff in the world. A piece of *arabushi* is as hard as a wooden chair leg, and one has to wonder to what use it can be put.

Arabushi is normally ground to a powder and much of it is used in the production of Hon-dashi, which is made by Ajinomoto. Hon-dashi is an important ingredient in many Japanese soup powders, prepared foods, and a variety of taste additives.

Arabushi can undergo yet more processing and be dehydrated further by fermenting it, resulting in a stronger taste. This is a time-consuming process and consequently the product, called *karebushi*, is more expensive and more sought after. There are actually two types of *karebushi*. One type uses the whole fillet and has a more bitter taste. The other type has a milder, more refined taste because the dark red bloodline of the fillet, which lies against the lateral side of the fish, is cut away.

Katsuobushi for sale at a Japanese market.

Unfortunately, we were not able to see how *karebushi* is made in the course of our visit to Yaizu. This is a description of the process. First the tar-covered outer layer of the *arabushi* is planed or scraped off and the fillets are sprayed with a mold culture (*Aspergillus glaucus*) at a temperature of 28°C. In the course of the following weeks, the mold spores sprout on the fillet and the fungal mycelia bore into the fillet. Once it is covered with mold, the fillet is brought out into the sun to dry, and all the mold is scraped off. This alternation between taking the fillet into the mold chamber and bringing it outside to dry in the sun continues for one to two months. The quality of the fish is thought to improve with each successive cycle.

The longer and more meticulous process used to make *karebushi* leads to a dried product that has fewer cracks. As a result, it is possible to make shavings that stay together instead of crumbling. When it is to be used, the hard fillet is placed on top of special box, which is like an inverted plane, so that ultrathin shavings can be cut from it. The *karebushi* yields the best taste when it is freshly shaved. Because they are so thin, it is technically possible to get 98 percent of the taste substance inosinate to seep out into a dashi. One can buy ready-cut shavings in airtight packages containing nitrogen to prevent the shavings from oxidizing. The shavings come in several varieties, according to the thickness of the shavings. Naturally, the thinner the shavings, the more quickly one can extract umami to make dashi.

Katsuobushi flakes are also sprinkled on soups, vegetables, and rice. When the dried flakes encounter the steam from the hot food, they contract and move as if dancing. That is why the Japanese call them 'dancing' fish flakes.

How does *katsuobushi* taste? First and foremost, there is a mild smoky taste, then a little saltiness, and then umami. Apart from inosinate, *katsuobushi* has at least forty other substances that contribute to its complex and distinctive taste, which combines saltiness, bitterness, and umami. Bitterness from an amino acid, histidine, is particularly prominent. The umami taste really comes to the forefront when *katsuobushi* is combined with other ingredients that contain glutamate, such as konbu.

▸ *Katsuobushi* production in Japan. Here the fillets are simmering in a special kettle.

KUSAYA

In Japanese, *kusaya* means 'rotten fish' or something that smells bad. Much like the Swedish *surströmming* and the Scandinavian *lutefisk*, however, *kusaya* is better than its name would have us believe. Even though it is truly malodorous, its taste is often mild. It is made from horse mackerel or flying fish, which are cleaned, deboned, and placed in brine. The brine is often reused time and again, and it is said that some families jealously guard their special brine for generations. The fish is kept in the solution for up to twenty-four hours, after which it is placed in the sun to ferment for several days to bring out its inherent umami. It should be possible to keep properly preserved *kusaya* for years.

Kusaya: small pieces of fermented, salted horse mackerel.

NORDIC VARIATIONS: HORRIBLE SMELLS AND HEAVENLY TASTES

Curing fish, such as herring, the old-fashioned way takes a long time. Whole, scaled herring, with their innards, are placed in layers of salt and cured in barrels that are kept for at least a month at 5°C. The enzymes from the stomachs and entrails of the fish help to break down the muscles and to release a reasonable amount of glutamate. This is why cured herring made by the traditional method have much more umami taste than those produced industrially. The factories employ a very rapid process in which the fish are placed for a couple of days in a strong vinegar solution, which tenderizes them by bursting their cells but releases only a little glutamate.

Just as in Asia, classical Nordic cuisines have their own varieties of fermented fish, which contribute umami to their diet. Particular examples are Swedish *surströmming*, Norwegian *rakfisk*, and Icelandic *hákarl*. All are notorious for giving off especially offensive odors, but many think them an essential, delicious feature of their national food culture.

Of these, *surströmming* is probably the one that should be considered the Nordic equivalent to *nare-zushi* and *kusaya*. It is made with small Baltic herring, which are caught in the spring, lightly salted, and placed in barrels to ferment for up to two months. They are then canned, but an anaerobic fermentation process continues due to the presence of lactic acid bacteria, which do not need oxygen. After a while, however, the carbon dioxide that is given off by the fermentation process causes the tins to bulge so that it looks as if they might explode. Although only a little salt is used, the fish do not rot because the fermentation is so robust. In addition, a series of sour substances, after which the dish is named, and free amino acids are formed.

Surströmming: fermented Baltic herring, a Swedish specialty.

Hákarl is an Icelandic specialty made from fermented shark.

Rakfisk, the Norwegian counterpart of *surströmming*, has been a national tradition for at least 700 years and possibly much longer. It is made from trout, whitefish, char, or, occasionally, perch that are caught in the early autumn. The entrails, blood, and gills are removed, and the fish are lightly salted and then placed in a vinegar solution for about half an hour. They are then placed in a barrel, with the open stomach turned upward and salted. A little sugar and some brine are added. The barrel is covered and the lid is weighed down lightly. The enzymes from the fish itself convert the proteins to free amino acids. Then lactic acid bacteria take over and a slow fermentation process proceeds at low temperatures, 4–8°C, in the course of the next three months. *Rakfisk* is usually served on flatbread or with potatoes and onions.

The original Nordic form of *gravlax*, which was fermented, is related to *rakfisk* and *surströmming*. In the Middle Ages, the fish was covered in flour and wrapped in birch bark, then buried and allowed to ferment. Modern *gravlax* now consists simply of fresh fish that is marinated for a few days in a mixture of sugar, salt, and spices. For this reason, there is not much change in the umami content of the fish.

The most infamous of the Nordic fermented fish is the Icelandic *hákarl*, made from either basking or Greenland sharks. Because it contains a great deal of trimethylamine oxide and also high levels of uric acid, the meat from these shark species can be consumed only after it is treated. In Iceland, the traditional method is based on fermentation. The fish is cleaned, the entrails removed, and the head discarded. It is then buried in a hole filled with gravel and covered with sand and more gravel. Heavy stones are put on top to weigh down the fish, so that the liquid seeps out.

The fish undergoes a fermentation process, in the course of which trimethylamine oxide, which has no smell, is converted to trimethylamine, which stinks horribly. At the same time, great quantities of ammonia, which has a pungent smell, are formed. After eight to twelve weeks, the shark is dug up and cut into strips that are air dried for two to four months. Before being served, the outer crust of the dried strips is removed and the cured fish is cut into cubes. *Hákarl* has a cheese-like, spongy consistency and a very strong smell of ammonia.

DRIED TUNA
Fillets of salted air-dried tuna are quite similar to *botargo*. Called *mojama*, they are of Phoenician origin and are now very popular in the tapas bars of Madrid.

Botargo: salted and dried roe from hake, ling, and gray mullet, a Mediterranean specialty.

FISH ROE

Roe contains a reasonable amount of glutamate, but how much it has does not always affect how strongly it tastes of umami. As evidence, we can cite the case of Russian caviar and salted salmon roe, which have 80 mg per 100 g and 20 mg per 100 g, respectively, of free glutamate. In theory, the caviar should have a much more pronounced umami taste than the salmon roe. But it is actually the reverse, because salmon roe also has inosinate, which is absent in caviar. The answer lies in the synergistic interaction between the inosinate and the glutamate.

BOTARGO IN THE DIARIES OF SAMUEL PEPYS, 1661
Botargo is mentioned in the diaries of the famous British author Samuel Pepys, written in London in the 1660s. On a very hot summer evening—June 5, 1661, to be precise—Samuel Pepys sat out in his garden with a friend, talking and singing, while drinking much wine and eating *botargo*. The salted roe had made him thirsty, and Pepys notes that when he finally went to bed at midnight, he was "very near fuddled."

▶ *Rakfisk*: Norwegian specialty made with salted fermented trout.

In Mediterranean cuisine, dried fish roe is a great delicacy, lending umami to such diverse dishes as tapas and pasta. The roe goes by a number of names: *botargo* in Spain, *bottarga* in Italy, *butàriga* in Sardinia, and *avgotaraho* in Greece. Roe sacs are extracted whole from tuna, swordfish, gray mullet, or a variety of fish from the cod family. They are put in sea salt for a few weeks and then hung up to air dry for about a month. The salt draws the liquid out of the roe, making it very firm and hard. A layer of beeswax, which is peeled off before eating, can protect the surface. Sometimes referred to as poor man's caviar, the dried roe is highly prized in Sicily and Sardinia and frequently served in Spanish tapas bars, where it is eaten with a little lemon juice. In Italy, *bottarga* is often cut into thin slices and used as a pasta topping, to add salt and umami.

Seven friends, *The Compleat Angler*, and a pike

The most famous book of all time concerning the art of catching fish is *The Compleat Angler*, written by the Englishman Izaak Walton (1593–1683). First published in 1653, it is written in the form of a dialogue between an avid angler and his traveling companions, a hunter and a fowler. Their conversation spans a journey of five days and gives a detailed account of how to catch and cook a variety of fish.

The pike fisher. Illustration in *The Compleat Angler*, Izaak Walton, 1653.

On the fourth day, the angler describes the pike, called by some "the Tyrant of the Rivers," and gives instructions on how to catch one using a frog as bait. He also reveals a secret recipe for roasting the fish, one that is loaded with umami from both the oysters and the anchovies that are stuffed into the body cavity. The former contains up to 1,200 mg free glutamate per 100 g, and the latter contains inosinate and melt right into the roasting fish. Walton's classic recipe is as follows:

First, open your Pike at the gills, and if need be, cut also a little slit towards the belly. Out of these, take his guts; and keep his liver, which you are to shred very small, with thyme, sweet marjoram, and a little winter savory; to these put some pickled oysters, and some anchovies, two or three; both these last whole, for the anchovies will melt, and the oysters should not; to these, you must add also a pound of sweet butter, which you are to mix with the herbs that are shred, and let them all be well salted. If the Pike be more than a yard long, then you may put into these herbs more than a pound, or if he be less, then less butter will suffice: These, being thus mixt, with a blade or two of mace, must be put into the Pike's belly; and then his belly so sewed up as to keep all the butter in his belly if it be possible; if not, then as much of it as you possibly can. But take not off the scales. Then you are to thrust the spit through his mouth, out at his tail. And then take four or five or six split sticks, or very thin laths, and a convenient quantity of tape or filleting; these laths are to be tied round about the Pike's body, from his head to his tail, and the tape tied somewhat thick, to prevent his breaking or falling off from the spit. Let him be roasted very leisurely; and often basted with claret wine, and anchovies, and butter, mixt together; and also with what moisture falls from him into the pan. When you have roasted him sufficiently, you are to hold under him, when you unwind or cut the tape that ties him, such a dish as you purpose to eat him out of; and let him fall into it with the sauce that is roasted in his belly; and by this means the Pike will be kept unbroken and complete. Then, to the sauce which was within, and also that sauce in the pan, you are to add a fit quantity of the best butter, and to squeeze the juice of three or four oranges. Lastly, you may either put it into the Pike, with the oysters, two cloves of garlick, and take it whole out, when the Pike is cut off the spit; or, to give the sauce a *haut goût*, let the dish into which you let the Pike fall be rubbed with it: The using or not using of this garlick is left to your discretion.

Walton concludes his tale by cautioning that "this dish of meat is too good for any but Anglers or very honest men; and I trust you will prove both, and therefore I have trusted you with this secret."

Klavs and Ole are both members of a rather peculiar little club, The Funen Society of Serious Fisheaters, who are all very familiar with *The Compleat Angler* and, of course, the story about the pike and how to cook it. Together with the other members, we set out to re-create the experience exactly as Walton describes it. We had been biding our time, anxiously, for more than a year, hoping that a pike of the appropriate size would eventually turn up in the cooler of our favorite fish dealer in Kerteminde, a nearby fishing town. One of the members, who is an enthusiastic sports fisher and a decorator, had been practicing how to tie together willow branches to make a sort of cage to hold the pike together as it was being roasted. In 2009, as we neared the beginning of the pike fishing season in May, when the fish is at its best, we hoped that luck might be with us. It was, and we soon received word that a pike that was absolutely ideal had been caught in a Swedish lake.

Right after the fish arrived in Kerteminde, on the afternoon before the roast, our willow cage maker hurried to the fish dealer's cooler to measure the fish. It was 120 centimeters long and weighed about 7 kilograms, a little bigger than the one yard recommended by Walton. He then set to work to prepare a cage that would fit the fish precisely. The willow twigs needed to be very green and thoroughly soaked in water so that they would not burn on the grill. The cage was tied together with wires and left open at both ends, which were closed up only after the pike was in place.

The next day, seven members of the club gathered at Klavs' restaurant, The Cattle Market, on the day of the feast. The pike was hung from the ceiling, cut open at the gills, and the entrails removed. Walton's recipe was followed to the letter. Fifty fresh oysters were stuffed into the cavity, together with a pound of butter, winter savory, and salted anchovies. Once it was completely full, a wooden skewer was poked through the fish to hold it together, and it was tied at the gills to prevent the stuffing from falling out. The "Tyrant of the Rivers" was placed ceremoniously in the willow cage and the ends were wired shut.

While this was going on, a charcoal fire was lit in the outdoor grill and, when it was ready, the fish was placed on it. The willow twigs and the fish were brushed with red wine to keep them moist. Walton specified that this should be claret, a red Bordeaux wine, which originally was a light rosé-like wine very popular in England in the 1600s.

After a little under an hour had elapsed, the pike was ready. The skin was still intact and none of the stuffing had escaped. The cage was cut apart and the fisheaters let the pike fall into a large platter, again as prescribed by Walton. The long-awaited moment arrived—the skin was peeled away and the pike cut into pieces. The flesh was almost translucent and easily slipped off the bones. We wondered how the stuffing would taste. There was no need to worry—it was wonderful. No liquid seeped out when the fish was cut up, as the butter had been absorbed right into the flesh and the oysters were cooked to perfection. The fish had a fantastic taste, with just the right degree of saltiness and so much umami that it seemed to fill the mouth completely. And, to our delight, the anchovies and the winter savory had drowned out some of the muddy taste that can be common in pike.

We ate every last morsel.

▸ An authentic pike roast, in accordance with Izaak Walton's recipe.

*I often think of my
Alsatian grandmother....
When I was little I used to love
spending time in the kitchen with her
and watching her work. She would
talk to me about what she was
doing. She was forever trying
to improve her recipes, to
add a little more of this,
a little less of that,
so that everyone
would be happy.*

*J'ai souvent pensé à ma grand-mère alsacienne ....
Ce qu'elle me donnait, c'est surtout d'y avoir passé
pour moi du temps en cuisine. Je la voyais faire; elle
ne cessait de perfectionner les préparations, d'y
ajouter ceci, d'éviter d'y mettre cela, de me dire ses
idées, de tenir compte des goûts de tous.*

Hervé This (1955–)

Umami from the land: Fungi and plants

As we turn our attention from the sources of umami that are found in the oceans to those that grow on land, we are struck by some important differences. While a great many marine organisms are excellent sources, the number of fungi and plants that would be described as having significant potential to contribute umami is more limited. On the other hand, some are able to supply both basal umami by way of free glutamate and synergistic umami from nucleotides, especially guanylate. And it is among the fungi and plants that we also find a few of the true umami superstars: shiitake mushrooms, fermented soybeans, tomatoes, and green tea.

UMAMI FROM THE PLANT KINGDOM
As fruits and vegetables mature, they develop a more intense taste. Ripe tomatoes and ripe green peas are especially good sources of glutamate, as are cabbage and mature potatoes that have been cooked. Of the plants that contain nucleotides, tomatoes and potatoes are again prominent and are joined on the list by green asparagus, lentils, and spinach. A particular plant extract, green tea, is also able to impart a strong umami taste.

Generally speaking, fruits have very little glutamate and are not able to contribute to umami in any significant way. Tomatoes are a spectacular exception to this statement, as their free glutamate content increases dramatically over time. When ripe, they have ten times more glutamate than when green. Without it, they taste tart, much like a tomatillo or a citrus fruit. As they ripen, their sugar substances and organic acids are also developing.

It is possible to enhance the umami contribution of some plant products by fermenting or pickling them. Examples include sauerkraut and kimchi,

Freshly baked rye bread with aged cheese and onions, accompanied by a wheat beer.

both made from fermented cabbage, and pickled gherkins, capers, and gingerroot. Normally, however, simply pickling vegetables with salt and vinegar will not release umami. What is required is fermentation.

Some nuts and seeds, such as almonds, sunflower seeds, and especially walnuts, are also good sources of umami. Soybeans, which are oilseeds, have a special relationship to umami. They are among our most protein-rich foodstuffs. They contain all the essential amino acids that humans need, but their glutamic acid is bound in the proteins. For this reason, all unfermented soy products, such as tofu and soy milk, have only a little umami. The reverse is true once they undergo fermentation. As we will see later, plant products made from fermented soybeans, especially in the form of soy sauce and hydrolyzed vegetable protein, are, from a global perspective, among the processed foods that are most widely used to impart umami.

Of the cereal grains, only maize (corn) can contribute free glutamate when it is in the unprocessed state. Rice, despite its large content of proteins with glutamic acid, has only insignificant quantities of free glutamate. The picture changes completely when it is brewed or fermented for purposes such as making sake or rice vinegar. The same is true for other grains, such as barley and wheat, which are used to make beer. Bread baked with cereal and flour products can also yield umami, particularly if the dough is allowed to rise slowly with the help of either yeast or sourdough. Yeast and lactic acid bacteria release the glutamate in the proteins of the cereals. ⋯> *Seriously old-fashioned sourdough rye bread*

SERIOUSLY OLD-FASHIONED SOURDOUGH RYE BREAD

Sourdough starter

300 g (10½ oz) rye flour, divided in two equal portions
1 tsp light brown sugar
¼ g (a pinch) active dry yeast
3 dL (1¾ c) soft water, divided in two equal portions
1 dL (⅖ c) plain yogurt

1. Mix together one portion of the rye flour, the sugar, yeast, one portion of the water, and the yogurt. Cover the mixture and let it stand in a really warm place for 24 hours.
2. Add the second portion of rye flour and the second portion of water, mix well, cover the dough, and let it stand in a warm place for another 24 hours. You should be able to smell a delicate yeasty aroma.

Sourdough rye bread

300 g (10½ oz) mixed whole grains (rye, barley, spelt, and/or buckwheat)
sourdough starter
800 g (1¾ lb) rye flour
150 g (5¼ oz) mixed bran (preferably including some chopped dried figs)
4–5 dL (1¾–2⅕ c) soft water
1 bottle (12 oz) unfiltered wheat beer
3 Tbsp dark malt extract
30 g (1 oz) sea salt

1. Soak the grains for 24 hours in plenty of water. Drain them well.
2. Mix the sourdough starter and all of the other ingredients in a bowl for 10–15 minutes or until the dough forms a soft ball. A standing mixer is almost indispensable, although this can be done by hand with a lot of muscle power. Pay attention to the water amount: Too little water will result in a dry bread and too much in one that is sticky.
3. Set aside about ½ L (1⅛ c) of the dough, place it in a covered container, and refrigerate. This will be your starter for the next time you bake.
4. Oil a 3 L (3 qt) bread pan with neutral-tasting oil; pour the rest of the dough in the pan.
5. Cover the bread pan with plastic wrap and refrigerate for 12–24 hours. Then leave the bread dough to rise at room temperature for 12–24 hours, depending on how warm it is in your kitchen. When the dough has risen to the rim of the pan, it is ready to be baked. Preheat the oven to 150°C (300°F).
6. Bake the rye bread for 2½ hours without the fan on, if your oven has one.

Sourdough bread made with white flour is wonderful when it is freshly baked and the crust is crisp. Day-old bread can still be used and find a key place in a real umami celebration. ⋯▷ *Anchovies, grilled onions, sourdough bread, pata negra ham, and mushrooms*

ANCHOVIES, GRILLED ONIONS, SOURDOUGH BREAD, *PATA NEGRA* HAM, AND MUSHROOMS

Serves 4

4 small new onions, with their roots and tops
extra-virgin olive oil
salt and freshly ground black pepper
100 g (3½ oz) small fresh button mushrooms
2 or 3 fresh chives, finely chopped
1 piece day-old sourdough bread
flaky sea salt
16 small anchovy fillets, preferably in olive oil
1 small piece Parmigiano-Reggiano
1 small piece frozen pata negra ham (jamón ibérico)
balsamic vinegar

If you cannot obtain *pata negra* ham, use the best air-dried ham you can purchase, preferably with a little fat still on it.

1. Clean and trim the onions, rinse them thoroughly, dry them, and coat them with a little olive oil, salt, and pepper.
2. Heat a grill or grill pan over high heat. Grill the whole onions for a short time.
3. Cut the mushrooms into thin slices and toss them with the chives and a few drops of oil.
4. Break the bread into coarse chunks, and sauté them in a bit of olive oil in a skillet until they are golden brown. Sprinkle with the flaky sea salt, and transfer to paper towels to absorb excess oil.
5. *To serve:* Arrange the bread chunks, anchovies, grilled onions, and mushrooms on plates with the shaved Parmesan, and grate the frozen ham over the dish. Finish with a few drops of balsamic vinegar and a drizzle of olive oil.

▸ Anchovies, grilled onions, sourdough bread, *pata negra* ham, and mushrooms.

DRIED FUNGI

When rehydrated, fungi are the raw ingredients that contain by far the most guanylate. (See the tables at the back of the book.) Shiitake and *matsutake* mushrooms, along with morels, top the list. Dried shiitake mushrooms also contain large amounts of free glutamate and hence can contribute both basal and synergistic umami. Yeast, which is also a fungus, plays a very special role, as the yeast cells contain a great deal of glutamic acid. If the individual yeast cells are split open, the cells' own enzymes can break down their proteins by hydrolysis, thereby releasing large quantities of free glutamate. This process is exploited in the industrial production of yeast extract, as we will see later. Yeast extracts are widely used in commercially processed foods to add umami.

> **MARIELA'S MARINATED FRESH MUSHROOMS**
> Fresh button mushrooms contain few umami compounds. But miso and soy sauce can turn them into delicious umami morsels using this easy recipe: Gently fry small button mushrooms in a little butter with a clove of finely chopped garlic. Remove from the heat when they are only just cooked and then marinate in a combination of soy sauce, *ponzu* sauce, and *shiro* miso. Let stand for at least an hour before serving.

Dried fungi are one of the most important non-animal sources of 5'-ribonucleotides, especially of guanylate, which synergizes with glutamate to produce a stronger umami taste. Generally, fresh fungi contain only a little guanylate, but in the course of the drying process free nucleotides develop, up to a level of 150 mg per 100 g. Also, the fungi that are darkest in color are normally able to impart more umami. Truffles are an exception to this pattern, as the white ones are richer sources than the black ones.

The shiitake mushroom is the classic example of a fungus in which guanylate is formed as it dries out. As mentioned earlier, they have a historical association with umami, in that the Buddhist monks in Japanese temples found out centuries ago that dried shiitake mushrooms could be substituted for *katsuobushi* to make a pure dashi without any animal products.

Shiitake mushrooms: dried fungi with a very high umami content.

Shiitake mushrooms are a very popular and highly prized ingredient, second only to button mushrooms as the most cultivated mushroom in the world. When fresh, they have a woody and slightly acidic taste, but as they dry they develop a wide range of taste and aroma substances, especially lenthionine, which is formed in the lamellae (gills) by the action of the mushrooms' own enzymes. This same process is repeated when the dried fungi are rehydrated in tepid water (30–40°C). In order not to weaken this enzymatic activity, it is important not to soak the mushrooms in water that is too warm or to cook or toast them.

There is a particular variety of highly prized shiitake mushrooms with small, dark caps called *donko*. These ripen over a long period of time and consequently develop more guanylate and are able to contribute more umami.

Other dried fungi that have a large guanylate content are morels, with 40 mg per 100 g, as well as porcini and oyster mushrooms, both with about 10 mg per 100 g. The Japanese mushrooms *matsutake* and *enokitake* are also good sources of umami.

FERMENTED SOYBEANS
Products made from fermented soybeans, in the form of liquids and pastes, are among the most widely used and oldest taste enhancers in the world and, in the form of protein-rich solids, they are ingredients in their own right in many prepared dishes. In addition to soy sauce, these products include miso, fermented tofu, and *nattō*, to name just a few.

Soybeans are probably native to Siberia or the northeastern part of China. Very early on in human history, the wild strains were domesticated and cultivation spread throughout East Asia. The oldest preserved samples of cultivars that resemble modern soybeans were found in Korea and date from about 1000 BCE.

There is clear evidence that soybeans were already being fermented in China more than 2,000 years ago. When one of the Han Tombs, sealed in about 165 BCE, was opened by archaeologists in 1972, black soybeans fermented with salt, *jiàng* (soybean paste), and *majiàng*, a type of fish paste with soybeans, were among the well-preserved foodstuffs found in pottery jars and identified by name on bamboo slips. Adding soybeans to fish that were being fermented to make fish sauce was a way to make the more costly ingredient stretch further.

SOY SAUCE

Soy sauce is found in the majority of food cultures in Southeast Asia. It is known by a variety of names, including *jiàng yóu* (China), *shōyu* (Japan), and *ganjang* (Korea).

The tradition of fermenting soybeans to make soy sauce probably arose in China, possibly at least 2,500 years ago. The spread of this practice is thought to have been driven by the necessity to limit the use of salt, which was expensive. Soy sauce could easily be used to add both savory and salty tastes to otherwise bland foods, such as cooked rice.

An eighth-century forerunner of Japanese soy sauce, called *hishio*, was produced from fermented soybeans to which rice, salt, and sake were added. This mass could be separated into a liquid and a more solid paste. The liquid was called tamari, which means 'accumulated liquid,' and it gave rise to *shōyu*, which can be translated along the lines of 'the oil or thick liquid from *hishio*.' The solid paste is equivalent to miso, an important component of vegetarian Japanese temple cuisine, *shōjin ryōri*. It is thought that the first commercial production of *shōyu* took place in Japan in 1290.

Modern *shōyu* is made from soybeans and wheat according to a process that has been in use in Japan since 1643. As wheat helps to impart sweetness to, and increase the alcohol content of, the end product, varying the proportion of these two ingredients results in a variety of taste nuances.

Soy sauce contains a great deal of free glutamate, with the actual amount being to a certain extent dependent on how the fermentation process is fine-tuned. A good Japanese *shōyu* has about 800 mg free glutamate per 100 g and Korean soy sauce somewhat more, as much as 1,264 mg per 100 g. The salt content is high, ranging from 14–18 percent.

Every country has its own traditional way of brewing soy sauce. In Japan and China, especially, there is a large selection of soy sauces, containing varying quantities of grain and characterized by differences in color, taste, salt content, and consistency. Chinese soy sauce has less wheat than the Japanese types, which are typically made from equal amounts of soybeans and wheat. Some cheaper brands based on hydrolyzed soybean protein are produced using fast industrial methods. This requires a shorter aging period and results in a different taste. A number of these products are actually sold as 'liquid aminos' and marketed as an alternative to naturally brewed soy sauce.

Brewing Japanese *shōyu* in the traditional manner takes time. Due to a shortage of soybeans in Japan after World War II, it was necessary to import them from the United States. Furthermore, because of the difficult economic situation at the end of the 1940s, the slow fermentation process, which breaks down the proteins in the beans to free amino acids, was displaced by a faster enzymatically controlled method. Later, there was a return to the classical methods, but in more efficient, modern breweries.

Commercially produced soy sauce.

> **PRODUCTION OF *SHŌYU***
>
> Brewing Japanese *shōyu* using the traditional process starts with a Chinese invention, a fermentation medium called *kōji*, which is the secret behind the whole process. Without it, it is not possible to make authentic soy sauce. For soy sauce, the *kōji* consists of a solid mass of cooked soybeans and roasted crushed wheat, which has been seeded with the spores of the fungi *Aspergillus oryzae* and *Aspergillus sojae*. The spores sprout to form a mycelium that grows in it for about three days at a temperature of 30°C. The resulting *kōji* contains a large quantity of hydrolytic enzymes that are able to break down proteins.
>
> The *kōji* is then placed in a saline solution (22–25 percent w/v NaCl) for six to eight months. The salt causes the fungus to die off, but its enzymes remain active. A yeast culture and lactic acid bacteria, which thrive in a salty environment, either start to grow spontaneously in the solution or are introduced into it. The combined activity of the enzymes and the microorganisms breaks down the proteins and fats, giving rise to a wide range of taste and aromatic substances, among them free amino acids and large quantities of glutamate. The resulting solid, called *moromi*, is placed in large, open cedar barrels, where it is allowed to cure for two summers. At the completion of this aging process, the *moromi* is transferred to canvas sacks that are pressed to separate out both soy sauce and soy oil. The latter floats to the top and is removed. The soy sauce is then pasteurized and bottled. Using this process, it takes two years to produce top-quality Japanese *shōyu*.

Shōyu was one of the few Japanese food products that the Dutch were allowed to take back to Europe during the 250-year Edo period, when Japan was isolated from the rest of the world by the shogunate. It was exported from their trading station at Dejima, near Nagasaki, and shipped to Europe, where it found favor in the kitchens of the royal French court.

Shōyu proved itself to be such an exceptional discovery that it has put its stamp on very many aspects of traditional Japanese cuisine. It is said that one reason why Chinese chefs assimilated globalized aspects of Western cooking so much more rapidly than the Japanese did, to put it simply, is that the latter were so enamored of *shōyu* that they did not set aside the extra time needed to evolve their cuisine.

MISO

The Japanese word for soybean paste is *miso*. It is thought to have originated at least two and a half thousand years ago in China, where it is called *jiàng*. Like soy sauce, miso is an offshoot from the practice of adding soybeans to fermenting fish and has evolved over time into a foodstuff in its own right. It is also possible that the forerunner of miso was not connected only with the fermentation of fish but also with that of shellfish and game.

Miso is produced from fermented soybeans together with different types of cereal grains. The process by which it is made is related to that used to brew *shōyu*. As a broad generalization, one can say that miso is the solid paste that remains after the liquids have been drained away from the fermented soybeans. Here again, the fermentation medium *kōji* plays a central role.

Miso: fermented soybean paste.

> **PRODUCTION OF MISO**
>
> The first step is making the *kōji*, following essentially the same method as the one for *shōyu*. If the end product is to be miso, however, either cooked barley or cooked rice is commonly used as a base, and it is seeded with the spores of the fungus *Aspergillus oryzae*. Soybeans are washed and cooked, allowed to cool, and crushed. Then *kōji*, salt, and water are mixed into the mush, which is placed in open cedar tubs and kept at a temperature of 30–40°C. This is the start of a slow fermentation period that traditionally lasts over two summers. During this extended period of time, the enzymes from the fungus, lactic acid bacteria, and yeast convert the soybeans and the grain from the *kōji* to the end product, miso. Typically, miso paste contains 14 percent protein and large quantities of free amino acids, especially glutamate. The salt content varies from 5–15 percent.
>
> Some commercially produced miso has a much shorter production cycle and derives its taste from artificial additives.

There are many varieties of miso, characterized by differences in color, taste, texture, and how long they have been fermented. Some varieties are *shiro* miso (white miso), which is very pale, has a mild, slightly sweet taste, and is often used in dressings and confections; *aka* miso (red miso), which is fermented over a long period of time, taking on a dark reddish color and acquiring a stronger, saltier taste; and *genmai* miso, made from brown rice. Subtleties of taste and color can also be attributed to the type of cereal grain that goes into the *kōji*. Although rice and barley are most common, many others, such as wheat, buckwheat, rye, and millet, can also be used either alone or in combination. Some types of miso are made with barley (*mugi* miso), and a very dark, expensive variety is made with soybeans only (*Hatchō* miso).

Finally, there are gradations in texture, depending on how finely the soybeans and cereal have been ground. In some cases, the soybeans have been mashed only coarsely, resulting in a consistency that is more like the original miso, which was made from whole beans.

DEEP-FRIED EGGPLANTS WITH MISO (*NASU DENGAKU*)

---

eggplants, preferably long, thin Japanese ones
neutral-tasting oil, for frying
miso
sesame seeds

1. Trim the stems from the eggplants. Cut them in half lengthwise.
2. Heat the oil in a deep, heavy pot over medium-high heat. Also preheat the oven broiler.
3. Deep-fry the eggplants until well done. Transfer to paper towels to cool slightly.
4. Score the cut surfaces in a diamond pattern and spread an even, thin layer of miso on them; sprinkle with sesame seeds.
5. Place the eggplants under the broiler and broil them until their surfaces are golden and slightly crisp. Serve warm. The eggplants are eaten by scooping out the inside with a spoon.

   The recipe can be tweaked to give it a Western twist as follows: Instead of sourcing the umami from miso, spread the eggplants with anchovy paste, possibly with a little finely chopped garlic mixed into it. Sprinkle yeast flakes and *panko* bread crumbs on top and broil until the crust is golden.

On account of its abundance of proteins, minerals, and vitamins, miso has been, and continues to be, an important nutritional resource in Japan. It is most often used to make miso soup. The soybean paste contributes additional umami to the soup, which is already loaded with savory taste substances from the dashi on which it is based. Miso can also be used to season oven-roasted or deep-fried vegetables. ⸱⸱⸱▷ *Deep-fried eggplants with miso (nasu dengaku) (page 115)*

A type of pickle, *miso-zuke*, can be made by placing vegetables (for example, daikon, marrow, and garlic) in a mixture of miso and sake. As part of the conservation process, some of the characteristic taste of the miso, especially umami, is transferred to the vegetables. The umami in vegetables with moderate amounts of glutamate (for example, white asparagus) can be intensified by combining them with miso and fish. ⸱⸱⸱▷ *White asparagus in miso with oysters, cucumber oil, and small fish*

WHITE ASPARAGUS IN MISO WITH OYSTERS, CUCUMBER OIL, AND SMALL FISH

Serves 4

4–6 thick white asparagus spears
1 dL (⅖ c) white miso paste
2 large firm cucumbers
4 oysters in their shells
4 Tbsp dried baby sardines
1 dL (⅖ c) neutral-tasting oil

1. Peel the asparagus carefully and place the spears in a sealed plastic bag with the miso paste; refrigerate for 3 days.
2. Peel the cucumbers to make about 50 g (1⅔ oz) peel. Place the peel in a thermo mixer or blender together with the oil.
3. If possible, blend at 70°C (160°F) for 8 minutes. Strain to remove any solid bits. The resulting liquid is cucumber oil, which can be kept in the refrigerator for several weeks.
4. Dry the marinated asparagus and cut them into thin slices lengthwise. Cut the slices again lengthwise, so that they look like noodles.
5. Open the oysters, drain their liquid into a little bowl, chop the oyster flesh coarsely, and mix it into the oyster water.
6. Mix the oysters with the asparagus, vacuum-pack the mixture, and refrigerate until ready to serve.
7. *To serve:* Arrange the asparagus and oysters in pasta bowls and pour any remaining liquid over them. Add a drizzle of cucumber oil and crumble the dried baby sardines over the top.

▸ White asparagus in miso with oysters, cucumber oil, and small fish.

THE ASIAN ANSWER TO CHEESE: FERMENTED SOYBEAN CAKES

Soybeans are also fermented to make protein-rich solid foods that are used as main ingredients in a variety of ways. Of the ones discussed here, tempeh and *nattō* are made from whole beans, while fermented tofu starts out as soy milk.

Tempeh, which comes from the island of Java in Indonesia, is one of the few soybean products that is not originally from China or Japan. It consists of whole soybeans that are soaked, hulled, partly cooked, and then seeded with a mold, *Rhizopas oryzae* or *Rhizopas oligosporus*. A little wine vinegar may also be added. The beans, which are still intact, are spread out in a layer and fermented for a couple of days at about 30°C. As a result, the mycelium from the fungus grows into the beans, so that they are more or less glued together into a cake but maintain the texture of the individual beans.

Tempeh: fermented soybeans; one type is made with seaweeds.

In the course of three days of fermentation, the free glutamate content in tempeh can skyrocket from about 10 to about 1,000 mg per 100 g, which, of course, dramatically increases its umami. Tempeh also contains a great deal of vitamin B. It has a mild but very complex taste with undertones of nuts and mushrooms and gives off a faint whiff of ammonia. In fact, the aroma is a little reminiscent of that given off by a good Brie at the perfect point of ripeness. Tempeh is often eaten as a replacement for meat, either cooked or fried.

A few years ago, a Swedish doctoral student, Charlotte Eklund-Jonsson, succeeded in creating an alternative version of tempeh by fermenting whole-grain barley and oats, plants that grow in colder climates. It is regarded as highly nutritious because, like its soybean counterpart, it has an abundance of folate. In addition, the method she developed was able both to preserve the high fiber content and to enhance the easily accessible iron in the final product, outcomes that are normally mutually exclusive.

Fermented tofu: *furu* and 'stinky' tofu.

Fermented soybean products are also made from tofu, which, like cheese, starts out as a liquid. Cheese is made from different types of milk, which contains lactic acid bacteria, by adding rennet, a complex of enzymes found in mammalian stomachs. The rennet causes the milk proteins to coagulate to form a very soft solid. This fresh cheese can be turned into a range of hard cheeses and soft ripened cheeses with the help of microorganisms and fungi, which break down the proteins to free amino acids. A similar process is used to make tofu from soy milk by adding a coagulant, typically calcium sulfate or magnesium chloride, to it. This causes the proteins to coagulate into a solid mass, which can then be pressed together to make it firmer.

The fresh tofu can be eaten in savory and sweet dishes or fermented to make other products. Just as cheese acquires a more intense taste as it ages, tofu takes on a stronger taste from a longer fermentation period.

Furu, a well-known type of fermented tofu from China, is related to tempeh. It is made from solid cubes of tofu that are air dried, fermented, and then immersed in brine to conserve them. *Furu* has 381 mg of glutamate per 100 g and is often used to add a savory taste to rice or vegetable dishes. Its texture can be compared to that of foie gras or a creamy blue cheese, and it has a mild, sweetish taste combined with saltiness from the brine. A special type of *furu* called 'stinky' tofu is fermented for more than half a year in a brine made with fermented milk and some combination of vegetables, herbs, meat, and dried seafood. While it gives off an incredibly strong, offensive odor, its taste is interesting and much milder. 'Stinky' tofu is an extremely popular snack food in Taiwan and Hong Kong, where it is usually sold by hawkers in the street and at night markets.

Tempeh and *furu* can quite correctly be regarded as the Asian answer to cheese. On occasion, *furu* has actually been called 'Chinese cheese.'

NATTŌ

The Japanese often use *nattō* as a little test to find out whether foreigners are really seriously interested in their culture. As with 'stinky' tofu, it can be a bit difficult to get past the odor and on to the taste. *Nattō* is made from fermented whole soybeans. It has a truly pungent smell, almost like that of an overripe cheese, and an intense umami taste. In addition, even though the beans are sort of glued together into a stringy, viscous mass, *nattō* is eaten with chopsticks. In Japan *nattō* is often eaten for breakfast, together with a raw egg on warm, cooked white rice. It is a very protein-rich food, with an abundance of vitamin K.

Nattō is made from small soybeans that are presoaked and then steamed for several hours before being seeded with *Bacillus subtilis nattō*, a bacterial culture. The soybeans are placed in a warm, moist environment for a day or so, allowing the bacteria to start a fermentation process. If tradition is being followed strictly, the container used for this is made from woven rice straws. After this short fermentation period, the soybeans are cooled to about 0°C and left in the cold for a week. During this time, enzymes break down the proteins to free amino acids, mainly glutamate.

The glutamate content of *nattō* is about 136 mg per 100 g, considerably less than in other fermented soybean products, simply because the fermentation is of such short duration. As no salt is added during fermentation, *nattō* is only slightly salty in comparison with many of these other products.

▸ *Nattō*: fermented whole soybeans.

Little is known about the origins of *nattō*, but it is likely that this food has been around for thousands of years. It is equally probable that the

tradition of fermenting whole soybeans with only a little salt arose independently in many places in Southeast Asia.

Other food cultures prepare a variety of beans in a similar way. Compared to *nattō*, some of these products have a much greater free glutamate content, for example, 1,700 mg per 100 g in West African *soumbala*, made from fermented *néré* (*Parkia biglobosa*) seeds, and Chinese *douchi*, made from salted soybeans, which turn black as a result of fermentation. The paste made from them is well known as black bean paste, often used in sauces for Chinese food.

Douchi: fermented black soybeans.

BLACK GARLIC

Black garlic is used as a food ingredient in Asia, particularly in Japan and Korea. It has recently put in an appearance at the cutting edge of Western cuisine. The cloves of the fresh garlic turn pitch black after being kept for several weeks at a temperature of 65–80°C in a closed container with a controlled humidity level (70–80 percent). Although often referred to as fermentation, the process involves enzymes and low-temperature Maillard reactions. During this process, the garlic softens and the rather pungent taste of the fresh garlic turns into a tangy, round, sweetish, and pleasant aromatic flavor with notes of balsamic vinegar and tamarind. In addition, the garlic has a prototypical taste of *kokumi* that is a delicious complement for dishes with umami.

Black garlic.

SHŌJIN RYŌRI: AN OLD TRADITION WITH A MODERN PRESENCE

The increase over time in the diversity of Japanese dishes made with the concentrated foodstuffs discussed in this and preceding chapters was undoubtedly related to major religious developments. Buddhism became the official state religion of Japan in the sixth century, and this resulted in stringent prohibitions against the consumption of meat and fresh fish, which lasted for hundreds of years. This gave rise to a need to find replacements that could introduce umami to make food more palatable. As a result, dried and fermented fish, seaweeds, vegetables, soybeans, and fungi were placed in the spotlight. Later on, when Zen Buddhism grew influential and the consumption of all fish was banned, it was no longer permitted to make dashi with *katsuobushi*. This was how the distinctive, strictly vegetarian Japanese version of temple cuisine, *shōjin ryōri*, began to take shape. In a sense, one can view the core of this cuisine as a search for deliciousness and umami without using any animal products. The only slightly less ascetic branches of modern vegan and vegetarian movements espouse these same goals.

GRILLED *SHŌJIN KABAYAKI*: 'FRIED EEL' MADE FROM LOTUS ROOT

120 g (4¼ oz) Japanese soy sauce
180 g (6¼ oz) mirin (sweet rice wine)
30 g (1 oz) sugar
300 g (10½ oz) fresh lotus root
rice vinegar

1 tsp katakuriko starch or cornstarch
½ sheet dried nori
vegetable oil
sanshō pepper and Manganji peppers
nama-shichimi

Serves 4

1. Prepare a thick sauce by mixing the soy sauce, mirin, and sugar and cooking it in a pot. Allow the sauce to cool.
2. Wash and peel the lotus root and soften it in water with some rice vinegar, about 1 tsp per liter (quart) of water.
3. Grate the lotus root finely and squeeze the water out of it with your hands.
4. If necessary, mix the grated lotus root with a little of the starch so that it has a reasonably stiff consistency.
5. Spread the starch on the sheet of dried nori and distribute the grated lotus root evenly over the surface of the whole sheet.
6. Preheat the oven broiler. Pour a little oil into a skillet, just enough to cover its surface, and turn on the heat.
7. Carefully place the lotus root-covered sheet in the skillet, nori side down; fry until the lotus root is golden brown.
8. Place the sheet with the lotus root in the oven and grill, brushing the browned lotus root several times with the thick sauce, until the surface is glossy and crisp.
9. Remove from the oven, sprinkle with the *sanshō* pepper, and cut into desired serving pieces.
10. Serve the *kabayaki* while it is still warm, possibly with crisply toasted pieces of lotus root and small green *Manganji* peppers stuffed with miso and *nama-shichimi*.

This centuries-old vegetarian temple cuisine, which is still practiced in Japan, has started to put in an appearance in Japanese restaurants in the Western world, where it is being infused with a new life and reintroduced in a modern context. Interest in *shōjin ryōri* is stimulated by the desire for simple, healthy food that is preeminently based on seasonal ingredients and in which palatability and umami are achieved by special combinations of fresh and prepared foods.

practice is known as 'the enlightened kitchen,' a reflection of the meaning of *shōjin* as refraining from evil, learning to do good, and awakening the mind. There is no effort to hide its roots in the ascetic Zen principles of simplicity, purity, harmony, and respect for nature. But *shōjin ryōri* is much more than just food for the Buddhist monks.

The owner of the prize-winning restaurant SAKI Bar & Food Emporium in London is a young, dynamic businesswoman by the name of Ayako Watanabe. She explained to me, Ole, that the enlightened kitchen is more than, and different from, purely vegetarian food. It is a concept that is underpinned by strict rules, which can be learned only over the course of many years. Furthermore, there are very few places in Japan where one can train to become a proper *shōjin* chef. She said that she had the good fortune to have been able to latch onto a newly trained chef, Yoshitaka Onozaki, from the best and most traditional *shōjin* restaurant in Japan, Hachinoki, in Kamakura, south of Tokyo.

At the time of my visit, Yoshitaka-san had been in London for about three years and was the head chef at SAKI. As he sat across from me with his hands folded, erect and with a serene expression, I could see clearly that for this chef making food was more than a job; it was a calling and a way of life. This was the point at which he had arrived after having completed two and a half years of

Chefs trained according to the precepts of *shōjin ryōri* are taught to draw out umami from purely vegetarian ingredients: vegetables, seaweeds, cereals, legumes, wild plants, and fungi. Creating delicious tastes in this type of temple cuisine presupposes that the chef has an intimate knowledge of the raw materials and how they impart umami when handled with both understanding and expertise. An example is lotus root, which has large quantities of glutamate, about 100 mg per 100 g. Another aspect of this cuisine revolves around the skillful use of dashi, the mother lode of umami. Other important ingredients are *shōyu*, miso, and *fu* (wheat protein). There is also great emphasis on what is succinctly referred to as *gomi*, *goshoku*, *gohō*, meaning 'five tastes, five colors, five ways' of preparing food. Meals, therefore, tend to feature ingredients with each of the basic tastes, having five different colors—namely red, white, green, yellow, and black—and being prepared in five different ways—either raw, boiled, baked, deep fried, or steamed.

SAKI specializes in serving *shōjin kaiseki*, a very high-level version of ordinary temple food. In contrast to the much simpler daily monastery fare, which includes only three dishes—a little soup, a bowl of rice, and an accompanying side dish—*kaiseki* consists of many. The Japanese expression for this elaborate meal embeds elements of respect (*kei*) and purity (*sei*), an indication that it was never intended for the monks, but rather for their important guests, often from aristocratic circles. *Kaiseki* is sometimes also served as part of the traditional tea ceremony (*cha-kaiseki*).

The secret at the heart of *shōjin ryōri* is an approach to cooking that is essentially the opposite of that of Western cuisine. In much of the latter, the flavor of a dish depends on adding spices and herbs in the course of preparation or at the very end, just before the dish is served. Many recipes sign off with the instruction 'season to taste.' In *shōjin ryōri*, the taste elements are introduced right at the beginning and almost always lead off with dashi, the soup stock that, as we know, is almost pure umami. It works its magic on vegetables and prepared products such as tofu, *yuba* (the skin from heated soybean milk), *fu*, cooked rice, and soups and is even incorporated into desserts.

Because temple cuisine cannot make use of fish, the *shōjin* dashi depends on fungi, particularly shiitake mushrooms, to interact synergistically with the konbu to impart an intense umami taste. The perfect gustatory balance is then achieved by adding dried daikon (Chinese radish), a little salt, soy sauce, mirin (sweet rice wine), *yuzu* (Japanese lemon), toasted tofu, and a special seven-spice combination. This spice is called *shichimi* and is a mixture of *sanshō* pepper, white and black sesame seeds, red chile, dried ginger, *ao-nori* (a green alga similar to sea lettuce), dried mandarin peel, and hemp seeds.

Ayako-san is convinced that the reason why the much-vaunted French nouvelle cuisine of the 1970s did not really catch on is that the combination of ingredients of exceptionally high quality and simplicity and minimalism in their preparation is, in itself, just not sufficient to ensure that the food is delicious. She feels that what was missing was the use of dashi from the very beginning. Even though the rules that govern the preparation of *shōjin ryōri* are very strict and steeped in traditions that go back for centuries, there is still room for innovation and renewal. Otherwise, things would become a little boring. She added that experimentation with new vegetables is permitted. In fact, it is natural and even necessary to use a whole range of non-traditional ingredients in order to adhere to the important idea of using what is locally available. A key element in applying this to vegetarian dishes is to tease out, possibly with the help of a little dashi, their characteristic tastes, which often change in the course of the growing season. And, of course, discovering new taste impressions along the way is always encouraged.

Grilled *shōjin kabayaki*: 'fried eel' made from lotus root.

It was also a revelation to learn from Ayako-san that the avoidance of meat and fish does not mean that one has to deny that one feels a need and desire to eat foods other than those that come from plants, algae, and fungi. Many *shōjin ryōri* dishes are actually named after meat and fish dishes, and serious effort goes into preparing them so that they will look the same. For example, they might resemble sushi and sashimi or mimic the appearance of fried eel. On top of that, it is absolutely acceptable for them to taste like meat or fish and to have a similar mouthfeel.   Grilled *shōjin kabayaki*: 'fried eel' made from lotus root (page 123)

TOMATOES

Even though tomatoes are actually fruits, we tend, from a culinary standpoint, to classify them as vegetables. On a global scale, they are the second most important vegetable crop after potatoes. Tomatoes are not just eaten raw, but they are also turned into purées, ketchup, juice, and salsas. Of all the vegetables, they have the greatest content of free glutamate, and it is enhanced when they are prepared in conjunction with other ingredients that contribute synergistic umami, for example, fish and shellfish. This is undoubtedly the reason why tomatoes are so popular and are used in an enormous range of dishes in kitchens all around the world. In fact, tomatoes also have a free nucleotide of their own, adenylate, rendering this popular food ingredient one of the few that can contribute umami synergy on its own.

It can be difficult to distinguish the taste of umami in a ripe tomato, even though free glutamate is one of its most important components. The sweetness of a ripe tomato easily overpowers the more subtle umami. Nevertheless, when it comes to vegetable sources of umami, tomatoes are the champions.

The free glutamate content of tomatoes increases more than tenfold as they ripen. (See the table at the back of the book.) Sun drying them increases the glutamate content even further. In addition, tomatoes also have fair amounts of the nucleotides adenylate and guanylate. This is a good reason to add ripe tomatoes to a green salad, which has little taste on its own.

A scientific article published in 2007 reported on an analysis of the glutamate and 5'-ribonucleotide content in thirteen different varieties of tomatoes. One of the authors is the well-known chef Heston Blumenthal, who has established one of the world's best restaurants, The Fat Duck, in England. He is a pioneer in the field of molecular gastronomy and also an enthusiastic advocate of umami.

The point of departure for the study was the chef's empirical knowledge that the umami taste of a tomato seems to depend on whether it is the outer flesh or the inner pulp, including seeds, that is being consumed. Surprisingly, the analysis showed that the glutamate content of the pulp is three to six times as great as that of the outer flesh and that the concentration of adenylate is at least four times as great in the pulp. This means that when preparing a dish with tomatoes, one must be sure to use the pulp and the seeds in order to maximize umami.

> **THIS IS UMAMI!**
> When Dr. Kumiko Ninomiya, one of the world's leading umami researchers, demonstrates what umami is, she uses a small sun-ripened tomato. She asks the person to take the whole tomato into the mouth and chew it about thirty times without swallowing. In the beginning, the tomato gives off a number of different kinds of tastes. But what remains at the end lingers and tapers off very slowly—*that* is umami!

MSG
126 mg/100 g
456 mg/100 g

AMP
8 mg/100 g
30 mg/100 g

Cutaway of a tomato showing the variations in the glutamate (MSG) and adenylate (AMP) content across the inside of the tomato.

As we learned from the experiment with making a Nordic dashi, sun-ripened tomatoes can be used as a source of glutamate for the stock, for example, in combination with smoked shrimp heads or dried fungi.

The wealth of umami in tomato juice can also be used to advantage in cocktails made with an alcohol that has little taste on its own, such as vodka. A classic example is the Bloody Mary. An even better example is a cocktail inspired by the sauce for a Venetian version of *spaghetti alle vongole*, made with clams and tomatoes. This is the Bloody Caesar, which combines vodka, tomato juice, clam broth, Worcestershire sauce, and a drop of Tabasco sauce. It is Canada's most popular mixed drink. But it is, possibly, also its best-kept secret and deserves a much wider following elsewhere.

Mackerel contains large quantities of inosinate, 215 mg per 100 g. When combined with tomato, the synergy between the two ingredients turns the dish into a veritable umami bomb. So it is little wonder that even a simple tin of mackerel in tomato sauce can give so much pleasure. And who has not experienced the fantastically good taste that the traditional Italian Bolognese sauce, combining tomatoes and meat, can add to a simple bowl of pasta?

Because tomato sauce enhances umami, it provides a convenient way to round out the taste impressions in a dish, both by increasing saltiness and sweetness and by masking any bitter nuances. ⇢ *Baked monkfish liver with raspberries and peanuts (page 128)*. The best way to gain the most concentrated umami taste from tomatoes is to oven-roast cooked whole tomatoes and herbs at low heat. If the tomatoes and herbs are chopped coarsely, the roasted tomato paste will end up with a texture that is similar to that of Bolognese sauce, but without the meat. The oven-roasted tomato paste can easily be substituted for a meat sauce in a pasta dish. ⋯> *Slow-roasted sauce with tomatoes, root vegetables, and herbs (page 130)*

Baked tomatoes can also enhance the taste of fried fish. When combined with sago pearls, which are basically pure starch but that can be made tasty by marinating them, the result is a dish with substantial umami synergy. ⋯> *Fried mullet with baked grape tomatoes, marinated sago pearls, and black garlic (page 132)*

BAKED MONKFISH LIVER WITH RASPBERRIES AND PEANUTS

Serves 4

200 g (7 oz) monkfish liver, presoaked in milk for 12 hours (see p. 58)
2 eggs
4–8 lovage leaves, chopped
50 g (1¾ oz) chopped shallots
2 anchovy fillets in oil

2 Tbsp puréed tomato
¼ garlic clove, chopped
a little sugar
pinch of cayenne
salt and freshly ground black pepper

Filling

2 Tbsp roasted salted peanuts

50 g (1¾ oz) fresh raspberries

Salad

¼ to ½ fennel bulb
75 g (2¾ oz) fresh raspberries
2 Tbsp salted peanuts

4 Tbsp raspberry balsamic vinegar
small croutons

1. Preheat the oven to 120°C (250°F). Blend the liver, eggs, lovage, shallots, anchovies, tomato puree, garlic, sugar, and cayenne, season well with salt and black pepper, put the mixture through a sieve to remove any lumps, and pour into 8 small molds greased with a little melted butter.
2. For the filling, lightly toast the peanuts in a dry skillet over medium heat. Distribute the peanuts and the raspberries among the molds.
3. Bake for 5–10 minutes. Check with a knife. The liver should be light pink.
4. Cut paper-thin slices from the fennel and arrange into a little salad with the raspberries and peanuts. Drizzle with vinegar, and sprinkle the croutons on top. Remove the monkfish liver from the molds and serve alongside the salad.

▶ Baked monkfish liver with raspberries and peanuts.

SLOW-ROASTED SAUCE WITH TOMATOES, ROOT VEGETABLES, AND HERBS

Serves 4

- *1 kg (2¼ lb) root vegetables, such as onions, celery, celeriac, parsley root (or parsnip), carrots, and leeks*
- *2 kg (4½ lb) very ripe fresh tomatoes, skins removed, or high-quality canned ones without skins*
- *4 cloves garlic*
- *1 dl (⅖ c) olive oil*
- *a lot of fresh thyme*
- *1 chile pepper*
- *50 g (1¾ oz) tomato paste*
- *4 or 5 bay leaves*
- *salt and freshly ground black pepper*
- *butter, for serving*
- *chopped fresh sage leaves, for serving*

1. Put the root vegetables, chile pepper, tomatoes, and garlic through a meat grinder. Warm the olive oil in a large pot and stir in the vegetable-tomato mixture.
2. Remove the big stalks from the thyme, chop it, and add it to the pot together with the tomato paste and the bay leaves.
3. Cover the pot with parchment paper and a tight-fitting lid and simmer on the stove top for 3–4 hours. Alternatively, bake in the oven at 140°C (290°F). The longer this cooks, the better.
4. Season to taste, divide into portions as desired, and store in glass jars in the refrigerator or freezer.
5. When the tomato sauce is to be used, warm it up to the boiling point and whip some butter into it—the more butter you add, the fuller the taste will be. Season with a handful of chopped sage.

TASTY VEGETARIAN PASTA WITH PENNE AND PARMESAN CHEESE
Cook a large portion of penne, toss with the slow-roasted sauce, and serve with freshly grated Parmesan cheese.

▸ Penne with Parmesan cheese and slow-roasted sauce with tomatoes, root vegetables, and herbs.

FRIED MULLET WITH BAKED GRAPE TOMATOES, MARINATED SAGO PEARLS, AND BLACK GARLIC

Serves 4

40 g (1⅓ oz) sago pearls
400 g (14 oz) ripe grape tomatoes
a few sprigs fresh thyme
1 clove garlic, sliced
½ dL (⅕ c) olive oil
1 shallot, finely chopped
a little sugar

Soy marinade
1 dL (⅖ c) soy sauce
1 dL (⅖ c) water
1 Tbsp fish sauce or garum
1 Tbsp sugar

pinch of salt
4 mullet fillets, with skin, weighing about 100 g (3½ oz) each
salt and freshly ground black pepper
2 cloves black (fermented) garlic, frozen

Yuzu marinade
3 Tbsp fish sauce
3 Tbsp yuzu juice
3 Tbsp sugar
a little finely chopped chile pepper

1. Bring 2 L (8½ cups) water to the boiling point, and pour in the sago pearls. Cover and allow them to cook for about 20 minutes. Drain the liquid and allow the sago to cool.
2. Mix the two marinades separately by combining the ingredients for each in separate bowls.
3. Divide the sago into two portions and place one portion in each marinade. Let them marinade for about a day. They can keep this way for about 2 days in the refrigerator. Preheat the oven to 100°C (210°F).
4. Place the tomatoes in a roasting pan with the thyme, garlic, and olive oil.
5. Sprinkle with the finely chopped shallot and a little sugar and salt. Cover with aluminum foil and bake for 20–30 minutes, until the skin on the tomatoes cracks.
6. Skin the tomatoes, place them in a small pot, warm them gently, and season as desired.
7. When ready to serve, season the mullet fillets with salt and pepper and fry them, skin-side down, in a skillet over medium heat. Be careful not to let them dry out.
8. *To serve:* Arrange the mullet fillets on plates with the baked tomatoes and the two types of cold sago pearls. Slice the black garlic very thinly and sprinkle over the top.

▶ Fried mullet with baked grape tomatoes, marinated sago pearls, and black garlic.

GREEN TEA

Extracts from green tea leaves have fairly large quantities of free glutamate, for example, about 450 mg per 100 g for those of the highest quality (*gyokuro*) and about half as much for the more ordinary varieties (*sencha*). On the other hand, roasted tea (*hōjicha*) has very little, about 22 mg per 100 g. But there is actually a completely different amino acid that is thought to be responsible for the strong umami taste in green tea.

Green tea leaves have very large quantities of the amino acid theanine, about 2,500 mg per 100 g. It is the most prominent of the taste substances in tea leaves, where it makes up about half of the free amino acids. Theanine is derived from glutamic acid and imparts umami.

Green tea is an important accompaniment to many types of Japanese food. The constituent elements of the meal are often taken into account in selecting the type and quality of tea to be served. For example, in Japan one might choose the very mild and delicate *gyokuro* to go with sweet dishes and bean confections and the more bitter, roasted *hōjicha* before or after a meal. *Maccha* is powdered green tea whipped up in warm water and served in a large drinking bowl. This is how green tea was originally made, and *maccha* is still the focal point of the Japanese tea ceremony.

As we will see later, green tea and *maccha* can also be used in sweet dishes to create a good balance in relation to bitterness and umami.

Maccha: whipped green Japanese powdered tea.

▸ *Gyokuro*: green Japanese tea of the highest quality.

*Everyone
eats and drinks,
yet only few appreciate
the taste of food.*

人莫不飲食也
鮮能知味也

Confucius (551–479 BCE)

Umami from land animals: Meat, eggs, and dairy products

In general, there are more free amino acids in the foodstuffs that are made from the organisms that grow in the earth than there are in those derived from the animals that live on it. On the other hand, animal-based foods are good sources of inosinate, which interacts synergistically to signal the presence of proteins. The umami content of meat and dairy products can be strengthened dramatically by preparing them in certain ways or by fermentation and curing. In particular, both simmering meat and bones over long periods of time and fermenting milk result in an abundance of umami.

THE ANIMAL KINGDOM DELIVERS UMAMI IN SPADES
Meat and dairy products are excellent sources of protein. The art of turning them into tasty food is, therefore, to a great extent a question of breaking down the proteins present in the raw ingredients. Animal innards generally contain more umami substances than meat from the animal muscles. A piece of liver, for example, foie gras, can also have a large quantity of glutathione, a compound that elicits *kokumi* and, thereby, also enhances umami. ⋯> *Mushrooms, foie gras, and mushroom essence (page 138)*

In order to release the umami, we employ a whole range of techniques: cooking, ripening, curing, drying, salting, smoking, and fermenting. These processes allow us to draw out free amino acids in greater or lesser quantities. Many researchers in the field of evolution posit that our distant ancestors began to cook their food, especially the meat, a few million years ago. Their early culinary efforts were a determining factor in the eventual evolution of *Homo sapiens*, whose brain required a substantial input of energy in order to be able to develop to such a large size. Cooking the raw ingredients was the only way to make them sufficiently nutritious.

Fresh meat is a major source of umami, as it contains both glutamate and the nucleotides that create synergy. Poultry meat, particularly from chicken, duck, and turkey, has a high concentration of glutamate, and the same is true for game and veal. Pork and chicken are rich in inosinate, while beef has a reasonable amount. As lamb is a poor source of umami, it benefits from being prepared in duck stock, which has a great deal of it.

The amount of umami in a piece of meat is proportional to how long it has been aged: the longer, the better. That is why beef is much tastier after it has been aged than when it is freshly slaughtered. It is also true that older animals are better sources of umami than younger ones. As we will learn about later, cooking, aging, curing, drying, salting, smoking, and fermenting, either alone or in combination, can appreciably intensify the umami taste in a meat product.

MUSHROOMS, FOIE GRAS, AND MUSHROOM ESSENCE

Serves 4

800 g (1¾ lb) fresh button mushrooms
1 L (4¼ c) chicken bouillon
1 sprig fresh thyme
1 bay leaf
1 lovage leaf

200 g (7 oz) small button mushrooms
100 g (3½ oz) fresh foie gras, frozen
1 Tbsp duck fat
flaky sea salt and freshly ground black pepper

1. Preheat the oven to 80°C (175°F). Place the first lot of mushrooms in a roasting dish with the chicken bouillon, thyme, bay leaf, and lovage.
2. Cover with aluminum foil and bake for 10–12 hours.
3. Remove from the oven, strain off the liquid, place it in a pot, and reduce it over high heat to an aromatic, intensely tasty essence. Discard these mushrooms.
4. Sauté the second lot of mushrooms in the duck fat in a skillet over medium heat. Season with sea salt and black pepper.
5. *To serve:* Place the mushrooms in deep bowls, pour the mushroom essence over them, and top with shavings of the frozen foie gras. Serve immediately.

▸ Mushrooms, foie gras, and mushroom essence.

Homo sapiens is a cook, and the enhancement of umami is the recipe

The culinary arts and their outcomes, in the form of good nutrition and especially umami, most probably go hand in hand with the evolution of humankind over the past two million years or so. The umami taste sends a clear signal that the food is nourishing and, consequently, important for survival.

According to Charles Darwin, the two principal forces that drive evolution are "natural selection and conditions of existence;" in other words, genetic inheritance and the environment. Of the two, Darwin thought that the environment has the greatest influence. This means that even though the genes, and the way in which they are expressed, control the biological function of an organism, the organism is limited by the possibilities afforded by its surroundings and by the laws of physics and chemistry. An important aspect of the environment is food, which in this way enters the picture as a driving force in the evolution of the species.

The available raw ingredients are one part of the equation. The other is the way in which, over time, humans have prepared and eaten them.

The Harvard anthropologist Richard Wrangham has studied the difficulties faced by our ancestors in securing enough food by hunting and gathering. A particular problem was to have sufficient time to prepare the raw ingredients so that they were readily digestible. Otherwise they might not provide the calories needed not only for physical activities but also to make energy available to create the conditions for the evolution of a large brain, which consumes 20–30 percent of the total energy at the disposal of the body as a whole. Wrangham's conclusion is crystal clear — the use of fire to cook food was the essential breakthrough. Modern humanity as we know it was able to develop in this way only because our ancestors learned to cook food, especially meat. By itself, raw meat is insufficient to fuel the body's needs, and that is why humans became cooks. It is striking that humans are the only living beings who use heat to prepare their food—this is a universal trait across all cultures and continents.

Cooking not only makes it safer to eat the food, but it also preserves it and makes it softer and easier to chew. As a bonus, the food tastes better, primarily because the cooking process releases free amino acids, resulting in umami. In an evolutionary context, it is especially important that heating made the food more digestible and resulted in more readily available nutritional substances and, hence, a greater calorie intake. Cooking gelatinates the carbohydrates in the starchy parts of plants and denatures the proteins in meats. As a consequence, humans need much less time to eat and digest their food than other animals. This gave us some distinct evolutionary advantages over those species that had to devote eight to ten hours a day to obtaining their food supply and possibly an equal amount of time to chewing it. For example, chimpanzees and other large apes currently spend more than six hours chewing every day!

Experiments have shown that we, as a species, could not survive on a diet of raw food, as it provides too little energy, leads to weight loss, and, most important, reduces reproductive capacity. And comparisons between human anatomy and that of other primates show that we have smaller teeth, weaker jaws, and a small mouth. For Wrangham, these factors are all signs that early in our evolution, possibly as long ago as two million years, we began to cook our food. Another characteristic that distinguishes us from the other primates is a smaller stomach and shorter intestine in relation to body size. This seems to be an indication that our digestive system has adapted to prepared food, consumed in smaller pieces, which can be converted in less time into more nutrients and more energy. Consequently, as the digestive system required less fuel to carry out its work, more calories were available for the muscles and the brain.

Broadly speaking, evolutionary biologists regard the period 1.8–1.9 million years ago as a time of transition from ape species (*Australopithecines*) to the hominins *Homo erectus*, who had a smaller brain and a lower brow than modern humans. It is presumed that an important driving force for this evolution was that, during this transition period, the ancestors of *Homo erectus*, the habilines, became hunter-gatherers and changed from herbivores into omnivores. *Homo sapiens* is thought to date back about 200,000 years.

While it is obvious that *Homo sapiens* is a cook, there is little certainty about when our ancestors started to use fire to prepare food. Wrangham makes the argument that this might have happened much earlier than is generally supposed. He singles out the small teeth, a smaller stomach, and a large brain compared to the body as evidence that *Homo erectus* had already begun to increase caloric intake not just by eating meat but also by cooking it.

Meanwhile, because of the culinary arts, humans have become more accustomed than other primates to the carcinogens that can be formed in food in the course of heating it. At the same time, we have become less resistant to the bacteria that are present in raw meat. In an evolutionary context, though, the advantages of having ready access to more calories have outweighed the disadvantages of the loss of vitamins and the possible formation of dangerous substances in cooked food.

Wrangham is also of the opinion that the culinary arts have been instrumental in the development of social structures and the division of labor according to gender. Men undertook the arduous, risky task of hunting, while women spent their days gathering fruits, plants, and roots. In addition, the women had to guard the food supply and have warm, plant-based meals ready when the men returned home. They also kept the pot going on the fire in case the hunt had been successful and there was something more to prepare.

Wrangham characterizes the 'marriage' between a man and a woman as an arrangement that was, from the very beginning, necessitated by our relationship to food. A man who had no woman at home to prepare warm, energy-filled food for him was in a bad way. He would have to chew on raw ingredients all night. The inevitable socializing associated with the eating of a warm meal was, at the same time, a stabilizing factor that fostered the creation of family and societal ties.

As a final point, Wrangham posits that long before the introduction of agriculture, the culinary arts freed humans to engage in activities other than foraging for food and eating it. We became truly human when we started to cook our food. And we recognize the food that is good and nutritious from its satisfying taste.

Umami was a key component in the evolutionary process that turned humans into cooks.

Milk and eggs from domestic fowl are particularly important animal products. Although cow's milk has only a little umami, dairy products such as cheese can have a great deal. Eggs are also a good source and are important ingredients in many prepared products; mayonnaise being a prime example.

PRESERVING MEATS IN THE TRADITIONAL WAYS

Perhaps it is just a coincidence, but all the techniques that are ideal for eliciting umami were discovered and perfected as methods for preserving foodstuffs that are quite perishable. This is particularly true for fish, but also for meat and milk products. While conserving them is simply a tool to ensure that they will keep, it often changes them beyond recognition. What is important is whether they are still edible and have kept, or even increased, their nutritional value.

FIRST YOU ADD SALT
Almost all preservation processes involve adding salt. For example, in the case of curing, fermenting, or smoking meat, the meat is initially salted to draw out liquid. At the same time, the salt inhibits the growth of undesirable bacteria and fungi.

Conservation of meat is often a race between competing processes—some help to ensure that the raw ingredients remain edible in one form or another, and others damage them or cause them to spoil. Air drying of ham is a good example. First the hams are salted lightly and hung up to dry. At the same time, a mold grows on the surface of the ham. A delicate balance involving the mold, the ambient moisture, and spoilage bacteria determines whether the ham will keep and be safe to eat.

Cooking, drying, salting, fermenting, and smoking are all techniques that humans have perfected over time to preserve foodstuffs so that they can survive being transported, keep from one season or even one year to the next, and not least remain edible and digestible. The question of whether or not they are palatable is often a judgment call that depends on getting used to how a particular item tastes. Many of these conservation techniques have made it possible to increase the umami content, but at the same time they have given rise to other substances for which one might need to acquire a taste. After all, not everyone takes to the taste of fermented fish or smoked meat right off the bat. And who immediately liked the taste of milk that had gone sour and been attacked by bacteria and molds? But without this transformation, there would be no yogurt or cheese.

Wherever meat is eaten as part of the diet, it is also preserved by air drying. This is a good way to maintain its nutritional value, and the dried meat can easily be stored and transported. The dehydration process is almost always carried out in conjunction with salting and sometimes smoking. The critical factor is to dry the meat sufficiently quickly to prevent the spoilage bacteria from taking over. An example is jerky, made from both domesticated and game animals. Beef jerky has a great deal of umami and keeps almost indefinitely. A South African version called *biltong* is also marinated in vinegar and often prepared with spices as well.

Ripening and curing of meat makes it softer and improves the taste, partly because this releases free amino acids and small peptides, as well as free nucleotides such as inosinate and guanylate. In chicken, the guanylate content peaks eight hours after slaughtering. Beef that has been allowed to hang for a while has more free glutamate than it does just after the animal has been killed. Nevertheless, not all substances that impart umami are enhanced by being aged for a longer period of time. For example, inosinate in beef and pork reaches its maximum value just one or two days after slaughter. (See the tables at the back of the book.)

AIR-DRIED HAMS

The practice of air drying hams is common in many parts of Europe, based on folk traditions that go back for centuries. They are both found as staple homemade products in country markets and sold as luxury goods of the highest quality. Three of the most famous ones come from the Iberian Peninsula, Italy, and France.

Jamón serrano is air-dried mountain ham made from the meat of Landrace white pigs and black-footed wild pigs. The most sought after is *pata negra*, also called *jamón iberico de bellota*, as it comes from free-range pigs that forage in the forests of Spain and Portugal. Because the pigs eat great quantities of acorns, the combination of fats in their meat mirrors the content of unsaturated fatty acids in the wild oak nuts. As a result, the fat is softer and the texture and mouthfeel of the ham are more pleasing.

To make *jamón serrano*, fresh hams are first cleaned and trimmed and then covered with coarse salt for two weeks to prevent attack by bacteria and spoilage and to draw liquid out of the meat. After they have cured, the hams are washed and hung up to dry for four to six months in a controlled environment with low temperature and high moisture. Finally, the hams are ripened for up to fourteen months under ambient conditions in special drying sheds in the mountains of Spain. During this time, mold grows on the surface of the hams, winning the competition against the bacteria that would spoil them. At the same time, the proteins in the meat are being hydrolyzed, forming vast quantities of free amino acids—an abundance of glutamic acid, as well as alanine, leucine, and glycine. (See the tables at the back of the book.) When fully ripened, the ham has about fifty times as much glutamate as it started out with and has become a true umami bomb. Sadly, as the end product is heavily dependent on the climatic conditions in these mountainous areas, it is difficult to duplicate the taste of authentic *jamón serrano* elsewhere.

Italy's legendary Parma ham (prosciutto *crudo*) has a history that can be traced back to the days of the early Roman Republic, long before the start of the Common Era. Only a little salt is used for the preparation of this delicacy, so it is sweeter than its Iberian cousin. After salting, the ham is coated with a layer of fat to prevent its being attacked by mold and bacteria and to allow it to dehydrate very slowly. During the ripening process, which takes at least one year, glutamate and other taste substances are formed.

An equally famous, salted air-dried ham is made in the area around Bayonne, in southwestern France. The fresh meat is coated with coarse salt and placed in a cool environment for twelve days before being dried in special temperature-controlled storage rooms for eight to nine months.

SALTED BEEF: PASTRAMI AND CORNED BEEF
The easiest and most traditional way to draw umami taste out of beef and veal is to boil both the meat and the bones to make a soup stock. But, just as for ham, there are a number of conservation techniques that enhance glutamate in beef. Two well-known examples are pastrami and corned beef. Both of these products have much more umami than raw beef in the form of a dish such as steak tartare. The practice of salting beef is not so much to preserve it as to make it tastier.

Pastrami is made from a cut of beef that has a fair amount of fat, often brisket. It is first cured in a brine to which spices, such as garlic, bay leaves, coriander, peppercorns, and mustard seeds, are added. After being allowed to dry slightly, it is either hot-smoked or cold-smoked. Next it is steamed in order to remove most of the smoky taste. During this final process, the connective tissue in the meat is broken down into gelatin, leaving it very tender.

When Romanian Jews immigrated to the New York area in the late 1800s, they brought with them their tradition of making smoked meat, and it soon became a popular deli favorite. It is generally served slightly warm on light rye bread, with coleslaw and a dill pickle on the side.

Salted air-dried sausages.

Like pastrami, corned beef is made from beef that is first cured for several days in a brine with spices. Sodium nitrite is usually added to the brine, as this is what gives the corned beef its distinctive pink color; otherwise it would look grayish. The meat is then removed from the brine and simmered until tender.

BACON AND SAUSAGES

Meat that has been smoked, especially bacon made from pork, contains much inosinate, which enhances the natural saltiness and umami found in eggs and, in addition, brings out their sweet taste. So it is little wonder that bacon and eggs have long been regarded as the perfect breakfast dish.

The combination of bacon and green peas is another good source of umami. Bacon can also be used to dampen the bitterness in cooked cabbage or in roasted game meats by enhancing their sweet taste.

Sausage making, especially from beef and pork, has a long tradition and a global reach. Initially, freshly made sausages have no more umami than the raw ingredients from which they are made, but when cooked they can become very tasty. Once they have been salted, dried, smoked, fermented, or cured, however, they develop much more umami and are often used to enhance the taste of other foods. This is particularly true for those made from pork.

Development of the craft of turning meat into sausages that were then preserved using one of these techniques was driven by the necessity to find ways to store meat. Dried cured sausages—for example, Spanish chorizo or Italian *salsiccia secca*, pepperoni, and salami—can be kept for very long periods of time without spoiling. That the sausages became good sources of umami was a bonus.

DAIRY PRODUCTS

Only modest amounts of free glutamate are found in dairy products, such as milk, butter, yogurt, crème fraîche, and fresh cheeses like bocconcini and ricotta. But once enzymes have gone to work on the milk to make aged and mature cheeses or blue cheeses, large quantities of free glutamate are released. Really hard cheeses, foremost among them Parmigiano-Reggiano, are the foodstuffs that yield up the most umami. Also, products made from goat's and sheep's milk have more umami than those made from cow's milk.

BLUE CHEESES

The history of blue cheeses goes back more than two thousand years; they were mentioned by Roman writers in antiquity. They are made in many countries, especially those in Europe, with the most famous probably being Roquefort from France, Gorgonzola from Italy, Stilton from England, and Danablu (Danish blue) from Denmark. They have very different consistencies; some are soft and creamy, while others are dryer and crumbly. What all blue cheeses have is common is that they are made using a *Penicillium roqueforti* culture.

As the cheeses age, there is enzymatic breakdown of the milk proteins to release vast quantities of free glutamate, for example 1,280 mg per 100 g in Roquefort. Danablu is also a good source, typically with 700 mg free glutamate per 100 g. During the aging period of ten days to eight months, the free glutamate content of Danablu increases more than thirtyfold.

On account of their strong, complex taste, and not the least also because of their rich umami taste, blue cheeses are very versatile. Apart from being appreciated on their own, they are often used to enhance the taste of other foods, such as dressings, sauces, dips, salads, pizzas, soufflés, and beef.

> **MARIELA'S BLUE CHEESE PATTIES**
> A very simple recipe based on blue cheese provides for umami synergy together with lean meat: Form extra-lean ground beef into medium-size patties with a piece of creamy blue cheese, about the size of a small whole walnut, flattened and embedded in the middle of each. Fry thoroughly in a skillet. The cheese completely melts into the patty, suffusing it with umami. Season to taste with a little sea salt.

▸ Stilton: English blue cheese.

AGED, DRIED, AND HARD CHEESES

The aging of cheeses is a complex process that involves a whole range of biochemical and physical changes. Enzymes from the rennet, lactic acid bacteria, and other microorganisms all come into the picture. They break down the milk protein casein into small peptides and free amino acids, especially leucine, valine, lysine, phenylalanine, and glutamic acid (glutamate). It is the taste of these amino acids that we identify with that of aged cheese. Because the microorganisms also need amino acids for their own growth during this period, the timeline for the free amino acid content of, for example, glutamate becomes very complicated, and it can peak and then start to decline.

Cheddar cheese is produced by heating the curds, salting and cutting them into cubes, and then draining them to remove the whey. As part of the aging process of cheddar cheese, different types of lactic acid bacteria are added to the curds. The ripening process produces great quantities of free amino acids, including glutamate. (See the tables at the back of the book.) The glutamate content increases more than tenfold in cheddar as it ages over an eight-month period.

This brings us to the special case of very hard cheeses. The general rule is, the harder and drier the cheese, the greater its free glutamate content. Of these, the superstar is undoubtedly Parmigiano-Reggiano, the much-loved Italian Parmesan with the thousand-year history. It is what is known as a hard, granular cheese and is made from a mixture of raw cow's milk and skim milk. After the curds have coagulated, they are broken up mechanically into pieces the size of grains of rice. Finished cheeses are enormous, weighing a whopping 38 kilograms.

Parmigiano-Reggiano is aged for more than two years, during which time the glutamate content can increase right up to 1,600 mg per 100 g. (See the tables at the back of the book.) The cheese is one of the processed foodstuffs that has the most glutamate, surpassed only by certain fish sauces. At the same time, because it is one of the few hard cheeses made with skim milk, it has a low fat content, and, as salt is used only on the rind, it is not very salty. So it is quite obvious why adding some Parmesan cheese is such an excellent way to improve the taste of other foods.

In addition to umami from glutamate, the other secret of Parmesan cheese is to be found in the small, crunchy crystals that one can discern in the cheese and that contribute to its mouthfeel. These crystals are not made up of ordinary salt, which is what many people think, but rather of calcium lactate and the bitter-tasting amino acid tyrosine.

Parmesan cheese is unrivalled in its ability to add depth to the taste of an otherwise insipid pasta or rice dish. Because it is so rich in umami, Parmesan can help to attenuate the bitterness in some dishes, especially if they also have a little sweetness that can be enhanced by the glutamate from the cheese. For example, in pesto sauce the cheese helps to soften some of the bitterness inherent in basil. Moreover, the dry rind, which can be saved and stored for a long time, is a good source for adding umami to soups and sauces.

Another use of Parmesan is in baking, for example, to add umami to breads or biscuits. A real umami treat can be created by mixing in some bacon at the same time, because it contributes inosinate, which interacts synergistically with the glutamate. To enrich the taste even more, one can add a few flakes of nutritional yeast to the dough. ⋯> *Parmesan biscuits with bacon and yeast flakes (page 150)*

It is possible to make a vegetarian version of the Parmesan biscuits by substituting the red alga dulse for the bacon. Dulse has a characteristic slightly smoky and meaty taste. No doubt that is why toasted dulse was once a popular snack in Scottish pubs. ⋯> *Parmesan biscuits with dulse (page 150)*

Parmesan cheese (Parmigiano-Reggiano).

Parmesan biscuits with bacon and yeast flakes.

PARMESAN BISCUITS WITH BACON AND YEAST FLAKES

40 g (1⅓ oz) bacon
80 g (2¾ oz) Parmigiano-Reggiano, finely grated
150 g (5¼ oz) all-purpose flour
15 g (½ oz) nutritional yeast flakes

25 g (1 oz) sugar
5 g (1 tsp) salt
pinch of cayenne pepper
80 g (2¾ oz) cold butter
possibly some bacon fat

1. Preheat the oven to 170°C (340°F) and lightly grease a baking sheet.
2. Cut the bacon into small, thin strips and fry it in a skillet until it is crisp. Allow it to cool. Reserve the fat and chill it.
3. Mix together the flour, yeast flakes, sugar, salt, and cayenne. Cut the cold butter into pieces and cut it into the mixture, along with a bit of bacon fat. Add the grated cheese and bacon pieces and knead it all together.
4. Form the dough into a roll, 2–3 cm (1½ in) in diameter, and refrigerate it for a while to firm it up. Then cut the roll into ½ cm (¼ in) slices.
5. Bake on the baking sheet until golden brown, 7–8 minutes.

PARMESAN BISCUITS WITH DULSE

The bacon in the recipe above can be replaced with the red seaweed dulse that itself has high levels of glutamate and hence umami. Simply skip instruction 2 and replace bacon with 15 g (½ oz) finely chopped dried dulse in instruction 3. In addition to imparting umami, the red dulse pieces show up as beautiful purple dots in the biscuits.

EGGS AND MAYONNAISE

Fresh eggs have just as much free glutamate as pork, but they have no nucleotides to speak of. Cooked eggs have more glutamate than fresh eggs. The glutamate content of many simple dishes is increased by the synergistic umami that eggs can bring to them. These include comfort foods such as bacon and eggs, an open-face sandwich with hard-boiled eggs and marinated herring, and a hamburger patty with a fried egg on top.

Mayonnaise has some umami because it is made with eggs. It is an emulsion of oil and vinegar that is stabilized with proteins and fats from the lecithin found in eggs. This type of emulsion has a good mouthfeel and mouthcoating, which elicit an enhanced perception of umami. In Japanese mayonnaise, umami is enriched even more by the addition of pure MSG. One can also augment it by adding a dash of Worcestershire sauce to the mayonnaise.

> **HARRY'S CRÈME FROM HARRY'S BAR**
>
> Harry's Bar in Venice is legendary. Even the story of how it was founded is the stuff of legend. When a dissolute, young American named Harry found himself too broke to order drinks at the hotel bar he frequented, the barman, Giuseppe Cipriani, loaned him the money to do so until he returned to the United States. He came back two years later, in 1931, and repaid his debt several times over, giving Cipriani enough money to start his own bar. He did just that and called it Harry's Bar. It has the distinction of having invented not just a famous cocktail, the Bellini, but also carpaccio. The bar became a magnet for the celebrities of the day, among them Charlie Chaplin, Orson Welles, Peggy Guggenheim, Alfred Hitchcock, and Ernest Hemingway, who immortalized it in one of his novels.
>
> Giuseppe Cipriani invented carpaccio, a dish of paper-thin slices of raw veal, in 1950 as the answer to a request from a lady whose doctor had put her on a strict diet of raw meat. As raw veal is rather bland, Cipriani concocted a special sauce, Harry's crème—a mayonnaise with a dash of Worcestershire sauce to add some umami—to drizzle over it. The brilliant shades of red meat with the contrasting white of the sauce reminded Cipriani of the colors in the paintings of the famous Venetian painter Vittore Carpaccio (1465–1525), and the dish was named accordingly.
> ⋯▸ *Harry's crème (page 152)*. ⋯▸ *Veal tartare with Harry's crème (page 152)*

HARRY'S CRÈME

1 egg yolk
1 tsp white wine vinegar
½ tsp mustard
a few drops of lemon juice
salt and freshly ground black pepper

2½ dL (1 cup) olive oil or canola oil
2 Tbsp heavy cream
reduction of Worcestershire sauce
a few drops of Tabasco sauce

1. All ingredients must be at room temperature. Prepare an ordinary mayonnaise by beating together the egg yolk with the vinegar, mustard, lemon juice, salt, and black pepper. Whip it until it has thickened a little. Whisk in the oil, a few drops at a time at first and then in a thin stream, until the mixture has the desired consistency.
2. Adjust the seasoning with the cream, the reduction of Worcestershire sauce, and possibly a few drops of Tabasco sauce and a little more lemon juice. The sauce should be thick and slightly piquant.

VEAL TARTARE WITH HARRY'S CRÈME

Serves 2

1 large piece fresh ginger, peeled
300 g (10 ½ oz) veal fillet or top round
canola oil
sea salt and freshly ground black pepper
broad beans (can be replaced with green peas or green asparagus)
1 shallot
20 medium shrimps
grated fresh horseradish
Harry's crème
dried dulse, for serving
watercress leaves, for serving

1. Process the fresh ginger in a juicer to produce 2 tablespoons juice. If you do not have a juicer, grate the ginger as finely as possible.
2. Mince the veal and mix it gently with enough of the ginger juice or grated ginger to give a somewhat sharp taste. Add the oil, salt, and black pepper, and mix until the taste is soft and spicy. Shape into 2 equal patties.
3. Use broad beans if in season. Shell them, blanch for 5 seconds in boiling water, and then plunge them briefly into ice water.
4. Thinly slice the shallot and put the slices in ice water until ready to serve.
5. Just before serving, peel the shrimps and marinate them briefly in the horseradish mixed with a bit of salt.
6. *To serve:* Spread a thin layer of Harry's crème on each plate, put a tartare patty on top of it, and arrange the shrimps on top of the tartare. Sprinkle the broad beans, pieces of dulse, shallot slices, and a few pieces of watercress over and around the meat.

▶ Veal tartare with Harry's crème.

*The greatest service
chemistry has rendered to
the science of food is the
discovery, or rather the
exact definition,
of osmazome.*

*Le plus grand service rendu
par la chimie à la science alimentaire
est la découverte ou plutôt la
précision de l'osmazôme.*

Jean Anthelme Brillat-Savarin (1755–1826)

Umami: The secret behind the humble soup stock

By far the majority of foodstuffs, when in their raw form, have little umami. Drawing out the fifth taste from them, therefore, becomes a question of breaking down their constituent parts, especially converting proteins into small peptides and free amino acids. Nucleic acids have little direct impact on nutrition, but they unquestionably have influence when it comes to taste and to interacting with glutamate to impart synergistic umami. This is especially true with the humble soup stock, a basic tool in virtually all cuisines. It seems that, since earliest times, cooks in all parts of the world have intuitively found ways to prepare it with ingredients that complement each other, with results that are rich in umami. Here we take a quick look at the chemical underpinnings of their 'soup science,' using some well-known stocks as examples.

SOUP IS UMAMI
Viewed through the lens of a gastro-chemist, soup stocks are textbook examples of successful experiments to bring out umami by combining raw ingredients that have glutamate and 5'-ribonucleotides. These are rarely both found in one raw ingredient. In order to have umami in abundance, it is, therefore, generally necessary to have two components, one that has basal umami and another one that interacts synergistically. It is equally important to take into account the multiplier effect of 1 + 1 = 8. Because of the synergy that is characteristic of umami, it often requires only small quantities of nucleotides to have a major effect on the taste of glutamate.

Traditional ways of making stocks and soups to achieve a delicious taste have evolved in the different cuisines all around the world. In France, fish or meat is cooked with vegetables such as carrots and celery; in Italy, fish and shellfish are combined with Parmesan cheese or ripe tomatoes; in Japan it is, of course, dashi, based on a combination of konbu and *katsuobushi*.

How the terms relating to soups are used varies somewhat across cultures, but there are some generally accepted definitions. A true stock is a thin, clear liquid based on lengthy simmering of meat, poultry, or fish with vegetables in water. It can also be called consommé or bouillon (from the French *bouilli*, meaning 'cooked'). A light stock is usually made from white meat and vegetables. For a dark stock, bones, meat, and vegetables are sautéed before the liquid is added. In a broth, small pieces of the raw ingredients are usually left in and egg or some starches, such as rice or barley, may be added to thicken it. At the other end of the scale is dashi, which is simply an extract and not boiled at all. Soups taste good only if they have a salt content of about 0.75 percent, but if the soup has a great deal of umami, this can be reduced by a half.

A simple chicken bouillon is an exceptionally versatile stock. ⋯> *Chicken bouillon (page 157)*. When a little of the chicken meat is added to it, together with a few herbs and some rice or noodles, it is transformed into a universal favorite.

The actual mixture of ingredients in the different soup stocks varies considerably. Japanese dashi is by far the cleanest tasting. In its finest and simplest form it is almost pure umami, derived from the synergy between glutamate and inosinate, from konbu and *katsuobushi*, respectively. Guanylate from shiitake mushrooms replaces the inosinate in the vegetarian version, *shōjin* dashi.

Shizuo Tsuji's classical cookbook, *Japanese Cooking: A Simple Art*, is as much a tribute to Japanese food culture as it is a cookbook. It is noteworthy that the word umami does not appear even once in this volume, which was first published in 1979. Nevertheless, Tsuji gives the fifth taste pride of place when he says that "Japanese cooking is deceptively simple. Its key ingredients are but two: a rather delicate stock (dashi) made from konbu (giant kelp) and flakes of dried bonito, and *shōyu*, Japanese soy sauce." As we now know, both dashi and *shōyu* are the very embodiments of umami.

Dashi is used in a long list of traditional Japanese dishes, not just to make a consommé (*suimono*) or a miso soup but also in a fondue (*nabe*), in simmered dishes (*nimono*), and as a dipping sauce for tempura.

Tsuji writes that there are, of course, many alternatives to dashi, but that without it, the dishes can only be considered to be *à la Japonaise*. Not surprisingly, he maintains that the authentic taste is lacking.

CHICKEN BOUILLON

---

1 chicken, preferably a free-range stewing chicken, 1½ kg (3 lb)
cold water
3–5 g (¾–1 tsp) salt per liter (quart) of water
10 peppercorns

Vegetables

4 celery stalks
4 leeks
3 carrots, peeled
3 parsley roots (or parsnips), peeled
2 onions

Bouquet garni

8 sprigs fresh parsley
8 sprigs fresh thyme
2 stalks lovage
bay leaves

1. Place the chicken in a large pot with only enough water to cover it; bring to a boil, and skim off the foam.
2. Reduce the heat to a simmer, and add the salt; as the liquid will be reduced, remember to use salt sparingly.
3. Add the peppercorns and all the vegetables. Tie the bouquet garni ingredients with kitchen string and add to the pot. Allow the bouillon to simmer in the covered pot.
4. From time to time, skim off the dark foam, but be careful not to remove the small amount of fat that floats to the top, as it is taste in concentrated form.
5. Allow the chicken to simmer until it is tender, then take it out of the pot, remove the meat from the bones, and reserve it for other uses.
6. Strain the bouillon and reduce the liquid to the desired concentration. It can also be reduced further to a very strong extract that can be used by the teaspoonful to add a little umami to a dish.

In traditional Western cuisines, but also, for example, in Chinese recipes, meat or bones, typically from beef or poultry, are almost always used. The umami content increases over time as the soup cooks, because a greater proportion of the proteins in the meat are broken down. As there are

Osmazome and *The Physiology of Taste*

The French lawyer and politician Jean Anthelme Brillat-Savarin (1755–1826) is considered one of the fathers of gastronomy. His most famous work, *The Physiology of Taste*, was published anonymously just two months before his death and has never been out of print. It can be described as a bit of a curiosity, much more a snapshot of his era than a true physiology text. In it, Brillat-Savarin takes on the role of spokesman for a protein-rich diet; his rejection of starchy foods is a little like that of the recent low-carbohydrate movement.

Brillat-Savarin had no knowledge of umami, but he had an intuitive understanding about something that was similar to it and from which all good soups derived their value. In his descriptions of meat and meat extracts, he makes reference to a substance he calls osmazome. It is said to be found in all meat and is an umbrella term for those taste substances that are soluble in cold water, in contrast to those extracts that result from cooking. Brillat-Savarin writes that it is the osmazome that lends a delicious, strong taste to soup.

It is Brillat-Savarin's contention that osmazome is more abundant in dark, red meat from mature animals and only slightly present in lamb, chicken, suckling pig, and even the white meat of larger fowl. Even though he treats osmazome as something relatively well defined and goes so far as to compare it with alcohol distilled from fermented wine, it is clear to us that anything this tasty must be a complex mixture of fats, proteins, salts, taste substances, and so on. Nevertheless, we must grant him that his description of osmazome is close to what we identify as proteins and free amino acids. As we know, they are a source of umami, even though the proteins in the meat are not necessarily dissolved in the water during cooking. We would also say that umami can come from substances that are extracted both in cold and in hot water.

Brillat-Savarin states that osmazome is responsible for the golden brown crust on meat when it is sautéed or roasted. He would not have known about Maillard reactions, only identified almost a century later. They cause carbohydrates and proteins to bind together when one roasts, sautés, or simmers meat. It is likely that the resulting delicious substances were partly responsible for his enthusiastic appreciation of osmazome in meat dishes.

The importance of osmazome is underscored by Brillat-Savarin's statement below that many a cook has been shown the door because he appropriated the first bouillon, presumably the tastiest, for his own use. He goes on to cite the case of a certain Canon Chevrier, who went so far as to invent covered stockpots that could be locked with a key in order to thwart such thievery:

It is also their prior knowledge of osmazome that has caused so many cooks to be dismissed, guilty of having appropriated the first bouillon. This was what made the reputation of the soupe des primes, *that led to the adoption of the practice of eating some broth with crusts of bread as a source of comfort while in the bath, and which led Canon Chevrier to invent stockpots that could be locked with a key.*

The idea of *soupe des primes* is itself fascinating. At first sight it may appear just to be the first stock, much like the first dashi, but it is actually considerably more complicated. An article in *Edinburgh Monthly* from 1820 comments extensively on the recently published *Tabella Cibaria*, a short Latin poem that is an elegant versification of a standard French bill of fare with copious additional notes and explanations. The author, Abbé M'Quin, explains that according to Rabelais there are two kinds of soup—*soupe de primes* and *soupe de lévriers* (soup fit only for the greyhounds). The *soupe de primes* was the first broth, which the young monks used to pinch from the kitchen, when they had a chance, on their way to the choir at the hour of the *Prime*, the service that took place around seven or eight in the morning. By then, the porridge pot (*porridge* is derived from the Latin for leeks, *Allium porrum*, which was a cherished ingredient in Roman soups) had been on the stove for a couple of hours in preparation for the midday meal at around eleven, and the soup had already taken on an interesting appearance and taste. "It was a sort of beef-tea, the lusciousness of which was enhanced by the pleasing idea of its being stolen" (M'Quin). Like the cooks to whom Brillat-Savarin refers, the novice monks were clearly not able to resist the temptation to pilfer a little soup from the stockpots many more taste substances in this type of stock than in dashi, the taste is not as clean and is more complex. To put it baldly, a Western or Chinese soup is made using a rather brutal cooking process that breaks down unprocessed raw ingredients, namely meat and vegetables, resulting in a broad spectrum of taste substances, not just umami. Japanese dashi is exactly the opposite, being a gentle extraction at lower temperatures from just two ingredients that have only a few taste substances. It is easy to make dashi, but it is based on products that have been carefully prepared and aged over an extended period of time. After all, *katsuobushi* is a much more advanced raw ingredient than a fresh carrot.

AMINO ACIDS IN SOUP STOCKS

It is characteristic of all soup stocks that they contain primarily glutamic acid and significantly smaller quantities of the other amino acids; hence, umami is the dominant taste impression. On the other hand, there is a significant variation in the distribution of the other amino acids among the different types of stocks.

A simple konbu dashi has, almost exclusively, glutamic acid and aspartic acid, the salts of which contribute umami. When *katsuobushi* is added, small quantities of other amino acids are introduced, among them alanine, which is sweet. The simple composition of these two Japanese stocks is the reason why they have a clean, understated taste that is the closest one can come to that of pure umami.

A REAL FIND: A DASHI BAR

A small shop called Soutatu is situated on a side street by the famous Nishiki food market in Kyoto. The shop has something truly unusual—a dashi bar. I (Ole) happened to discover this unique gem when I was rushing away from shopping at the food market. Out of the corner of my eye I registered a display stand with some konbu in the window of the shop. I wheeled around to find out what this could possibly be. It turned out to be an absolutely wonderful place with a counter similar to one in a bar where the customers could sample various kinds of dashi. Can you imagine a place that focuses only on the pure taste of umami and a shop that only sells konbu, *katsuobushi*, and packages of correctly matched pairings of these two ingredients, ready to be extracted in water to make the most exquisite dashi?

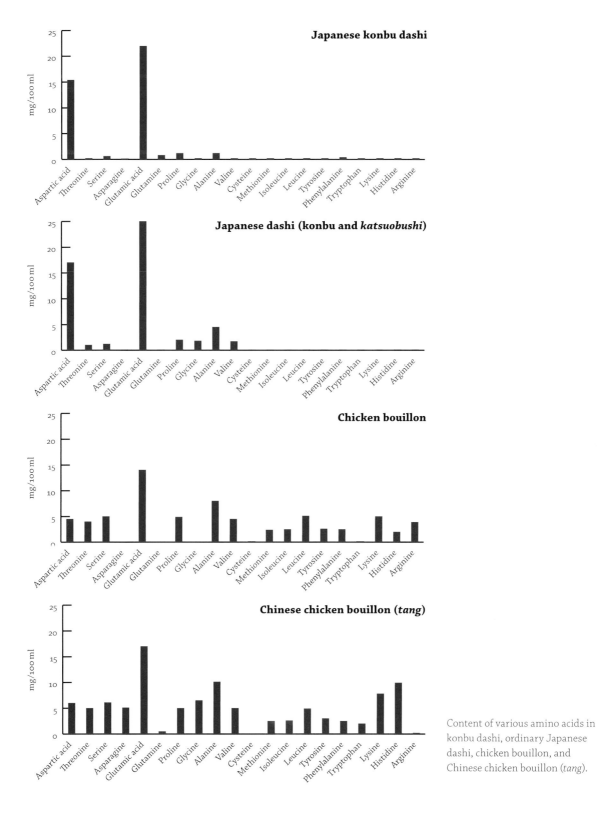

Content of various amino acids in konbu dashi, ordinary Japanese dashi, chicken bouillon, and Chinese chicken bouillon (*tang*).

Umami: The secret behind the humble soup stock

The type of chicken stock made in Western cuisines brings many more elements into play. Glutamic acid still predominates, but in addition there is a broad spectrum of other amino acids, with both sweet and bitter tastes. Consequently, the taste is much more complex than that of dashi.

We can also make a comparison with another typical Asian soup stock, Chinese *tang*, made by cooking crushed chicken bones, sometimes with pork as well. Again, glutamic acid predominates, but there is also a fairly even distribution of other amino acids that introduce taste nuances other than umami. Interestingly, the two most legendary Chinese soups, shark's fin soup and bird's nest soup, derive all their taste from the other ingredients, as both shark's fins and bird's nests are rather bland.

Meat or poultry in combination with vegetables imparts a great deal of umami. The taste is enhanced if the vegetables and bones are browned before they are simmered. Meat and poultry contain both glutamate and nucleotides, resulting in synergy. It is worth noting for future reference that both chicken and duck are richer in glutamate than beef and pork.

THE TASTE OF A BEEF STOCK

The classic soup stock in Western European cuisine is made with beef and beef bones. Obtaining maximum taste depends on cooking, rather than simply extracting them in liquid. Isabel Cambero and her colleagues at the Universidad Complutense in Madrid, Spain, have studied how beef broth flavor is developed and found that cooking temperature plays a more important role than cooking time. Apart from the free amino acids that are formed and their chemical reactions with sugars (by way of Maillard reactions), the relative concentration of inosinate is what gives the stock its characteristic taste, which is redolent of umami. Changing the cooking temperature from 55°C to 95°C releases three times as much inosinate. In addition to glutamate and aspartate, the complex flavor of the stock depends on the presence of both sweet (lysine) and bitter (methionine) free amino acids. The interplay among these substances results in a stock that is much more complex than dashi.

A good fish stock can be made using only fish and shellfish, as they contain both the glutamate and nucleotides needed to impart umami. The combination of shellfish with vegetables that have a good quantity of glutamate yields a particularly robust stock with intense umami. A good example is green pea soup with scallops, where the green peas contribute glutamate and the scallops release large quantities of adenylate.

⇢ *Green pea soup with scallops and seaweed*

Green pea soup with scallops and seaweed.

GREEN PEA SOUP WITH SCALLOPS AND SEAWEED

---

2 large onions
3 cloves garlic
neutral-tasting oil
1 kg (2¼ lb) small green peas, fresh or frozen
4 dL (1¾ c) water
2 dL (⅞ c) whole milk

3 Tbsp aromatic vegetable sea salt (e.g., Herbamare)
salt, seaweed salt, and freshly ground black pepper
6 large scallops
dried sugar kelp (optional)

Serves 6

1. Chop the onions and garlic coarsely and sauté them in hot oil until they are almost transparent.
2. Set aside 3 Tbsp of the green peas for adding to the soup later.
3. Add the water, milk, and Herbamare to the pot and cook for 5 minutes.
4. Mix in the remaining green peas and bring to a boil.
5. Mix everything in a blender, and season with salt, seaweed salt, and pepper. Add more milk if required to achieve the desired consistency.
6. Mix in the reserved peas.
7. In a skillet over medium-high heat, quickly sear the scallops on both sides without any fat until they are barely cooked through.
8. Serve the soup in bowls with a scallop placed in each, and sprinkle crumbled sugar kelp on top.

Purely vegetarian soups can derive glutamate from a variety of sources, such as cooked potatoes, green peas, soybeans, corn, green asparagus, cauliflower, mushrooms, and, of course, tomatoes. Fungi, especially dried shiitake mushrooms, have good quantities of the nucleotides that can interact synergistically, but there are very few vegetables that can play this role.

READY-MADE UMAMI

It is quite possible to take a shortcut and make a good soup with a great deal of umami by using one of the many ready-made vegetable-based bouillons that are available in liquid, solid, and powdered form. The first mass-produced commercial product of this type in Europe was made in about 1870 by Knorr, but the one that has become iconic is the famous Maggi cube.

No matter whether one starts with basic raw ingredients to make a stock or works with an elaborately prepared foodstuff like *katsuobushi* or the readily available concentrates and powders for Asian and Western soups, one is relying on identical principles for enhancing taste. One might say that a soup is where the umami orchestra plays a whole symphony, often with perfect synergy between substances that complement each other.

A FAST UMAMI KICK IN THE LUNCH BAG

In Scandinavian countries, cod roe in small cans is a popular item in children's lunch bags. They like cod roe just as much as mackerel in tomato sauce. Its popularity is no doubt due to the umami in the roe, which is enhanced when a little tomato paste has been added to it. Strangely enough, there is not much that smells worse than cod roe or mackerel in tomato sauce in a lunch bag, but it still has a wonderful umami taste once it is in one's mouth.

A gastronomic interpretation of an old umami classic: mackerel in tomato sauce; here fried mackerel is presented in a can with sun-ripened tomatoes, morels, charred potato, white asparagus, and *garum*.

KNORR AND MAGGI: EUROPEAN UMAMI PIONEERS

Carl Heinrich Theodor Knorr (1800–1875) was a German food products manufacturer, who started out by producing coffee substitutes and marketing grains. About 1870, with the collaboration of his sons, he introduced a line of soup powders made from finely milled dried legumes and dried vegetables. The company is now owned by Unilever, which has the rights to the Knorr brand in all countries except Japan, where it is marketed by Ajinomoto.

Knorr's soup powders and bouillon cubes were the first successful commercially produced soup concentrates in Europe. Among industrially prepared stocks, they are the ones that most closely resemble Japanese dashi powder.

The Swiss miller and industrial food pioneer Julius Maggi also came up with the idea of producing soup powders for the European market. They contain dehydrated, concentrated bouillon based on hydrolyzed plant protein, which is rich in glutamate—in other words, concentrated umami. His most famous, and ingenious, invention is the little bouillon cube, which was first sold in 1908. This inexpensive cube was used by those who could not afford meat, in order to impart a meaty taste to soups and sauces.

Even though Maggi's company was swallowed up many decades ago by the giant conglomerate Nestlé, the Maggi cube is still synonymous with a quick, cheap way to make a delicious bowl of soup.

A related product is a thin sauce, also called Maggi, which is popular in both the East and the West as a condiment or as a seasoning for soups and stews. Its taste is very similar to that of the herb lovage, which is not actually a component of the product. Ironically, however, lovage is referred to in a number of languages as the Maggi herb.

*The discovery of
a new dish does more
for the happiness of the
human race than
the discovery
of a star.*

*La découverte d'un mets nouveau
fait plus pour le bonheur du genre humain
que la découverte d'une étoile.*

Jean Anthelme Brillat-Savarin (1755–1826)

Making the most of umami

The time has now come to take a little tour of the various ways in which we can tease out umami in our own kitchens. To do so, we have included a number of recipes for both traditional and modern dishes that are made from selected raw ingredients treated in such a way as to take advantage of every last bit of umami in them. First, though, we will make a quick detour to discover some easy tricks for enhancing the taste of foodstuffs that may not be very savory on their own by adding certain readily available products to the dish.

MSG AS A FOOD ADDITIVE
The simplest way to impart umami is by adding MSG, which is sometimes marketed as 'the third spice', the other two being salt and pepper. In its pure form, it is incorporated in many industrially prepared foods to improve their taste.

Many studies have been undertaken to determine how the tastes of different raw ingredients and dishes are affected by the addition of MSG powder or crystals. These have been inconclusive in a number of cases, with no agreement about the outcomes. Nevertheless, we can outline some common general characteristics.

Here is an overview of how MSG can be used. The taste of both raw and cooked vegetables can be enhanced by the addition of very small quantities of MSG (less than 0.5 percent); it has a pronounced effect on carrots and cauliflower. MSG can also lessen the unpleasant taste impressions made by completely raw meat and some raw vegetables, especially an earthy taste in the latter. It brings out savory tastes in dishes with meat, fish, and shellfish, as well as in soups and stews. It would also appear that MSG can improve the taste of foods that have been frozen—in particular, oily fish, ham, and pork sausages. In foodstuffs with a certain sugar content, MSG and umami can help to lessen the taste of bitter substances, if any are present. On the other hand, there is no point in putting MSG on fruits or in fruit juices, sweet baked goods, dairy products, and prepared cereal products, as it does not alter their taste.

There is little doubt that the bad reputation attached to the use of MSG can be attributed to the rapid industrialization of food production that really took off in the 1960s. It was used to help companies turn out products of highly variable quality, where lack of taste was due to ingredients of inferior quality or overly rough handling to preserve them. More recently, the obsession with adopting a leaner diet has led to our banishing from the dinner table many fats, which contribute greatly to overall deliciousness and mouthfeel. One way to compensate for their loss is to add more umami, and here MSG can play a role.

How MSG is listed on food labels and whether it is deemed to be a food additive or a taste enhancer varies from one country to another. Given that umami derived from MSG has now gained recognition as a basic taste, however, it seems a little misleading to single it out in this way. Based on the extensive research that has been carried out on monosodium glutamate, it is one of the additives granted the status of "generally recognized as safe" by the United States Food and Drug Administration. Similarly, the European Union, the United Nations Food and Agricultural Organization, and the World Health Organization have all declared MSG to be harmless under normal circumstances and when used properly for its intended purpose.

OTHER COMMERCIAL SOURCES OF UMAMI
Glutamic acid and its other salts in their pure form are considered by most regulatory agencies to be food additives and must be listed as such. These rules also apply to sources of synergistic umami, the nucleotides, such as sodium inosinate, potassium inosinate, and calcium inosinate.

There is, however, a long list of substances other than pure MSG that are sources of umami. Faced with a growing movement on the part of many consumers to avoid foods that incorporate any of these so-called chemicals, manufacturers have had to find other ways to add glutamate. One solution is to incorporate appropriately treated protein-rich ingredients that are classified as food rather than as food additives. Often these are in the form of what are known as hydrolyzed proteins, from either animals or plants, and yeast extracts, all of which contain considerable quantities of free glutamate. Depending on the jurisdiction, these may be identified by their common or usual names.

The most widely used of these protein additives are derived from plant material with an abundance of glutamate, such as soybeans, and from yeast. For vegetables, the usual process is hydrolysis, which can be carried out with the help of a strong acid to release glutamic acid and glutamate from their proteins. Yeast is broken down by autolysis, a process in which the enzymes found in the yeast itself are used to rupture the cell walls. The resultant product is called yeast extract.

There are, in fact, quite a few well-established products, introduced long before detailed food labeling became mandatory, that depend on hydrolyzed proteins and yeast extracts for their intense umami tastes. Some of the best known are Marmite from Britain, Vegemite from Australia, Maggi sauce from Switzerland, soy sauce, and various fish sauces. They are all full of free glutamate, but as they are made from natural ingredients, its presence is not explicitly spelled out in their nutritional information.

HYDROLYZED PROTEIN

Proteins can be broken down with the assistance of enzymes, which cleave the proteins into smaller entities such as peptides and free amino acids. The process is called hydrolysis because it involves the use of water, the H_2O molecules of which are split apart in the course of the reaction. A strong, inorganic acid is often used as a catalyst and then recovered after the hydrolysis has run its course by neutralizing it with a strong base. Raw ingredients sourced from both vegetables and animals are hydrolyzed to release free glutamate.

The most commonly used sources of hydrolyzed vegetable protein are soybeans, corn, and wheat. Animal hydrolyzed protein is sometimes made from milk protein (casein) or from whey proteins derived from cheese production.

Industrially produced soy sauce is, fundamentally, a solution of hydrolyzed vegetable proteins.

The hydrolyzed protein products are purified and typically end up with a 5–20 percent content of free amino acids, for example, in the form of MSG.

Umami in a jar: Portable, affordable, and nutritious

The imperative of providing enough nutrition and tasty food for the population as a whole left its mark on the early industrialization of food production, directing its attention toward taste enhancers, with umami at the forefront. The three products described here are prime examples of products that were invented to answer these needs and are still popular many decades later.

In the 1870s, John Lawson Johnston, a Scotsman living at the time in Canada, developed one of the first mass-produced taste enhancers, a beef extract with a strong, salty taste. He came up with the idea in response to the contract he had been awarded to procure preserved meat for Napoleon III's army, whose troops could not fight on an empty stomach during the Franco-Prussian War. But as Johnston was unable to secure a sufficient supply of meat, he invented a sort of 'liquid beef,' which he called Bovril, a name derived partly from the Latin word for beef (*bos*). It is sold to this day as a thick, dark paste. It can be spread on bread, used in soups, or diluted with boiling water to make a hot drink. Bovril was eaten by British troops during both the Boer War and World War II. It has also proven to be a boon for civilians. For more than a hundred years, British and Scottish soccer fans have warded off the cold in the stadiums by bringing along thermal flasks filled with hot Bovril beef tea. And it secured its place in history from the reports in the *New York Times* that the men of Ernest Shackleton's 1916 expedition to the Antarctic fortified themselves with steaming cups of Bovril when they were marooned on Elephant Island and almost perished.

Advertisement for Bovril dating from 1907.

Unilever, which now owns this product, stopped using beef as the base in 2004 and started making it with yeast instead. It is quite likely that this change coincided with the fear prevalent at the time of contracting mad cow disease. Nevertheless, the manufacturer switched back a few years later, although the variety made from yeast extracts, suitable for vegans and vegetarians, remains available.

In Western countries, the best-known and currently most widely used specialty food with an abundance of umami is probably Marmite, which first saw the light of day in 1902. In some senses, Marmite can be considered a parallel product to Bovril. It was developed at Burton upon Trent, the site of what was then England's largest brewery,

Bass Brewery, and a source of large quantities of yeast, a by-product of the beer-making process. The famous German chemist Justus von Liebig (1803–1873) had earlier discovered that brewer's yeast could be converted into a useful, nutritious substance. Based on his findings, the Gilmour family set up a factory to take advantage of the excess brewer's yeast that would otherwise have been discarded. The yeast is hydrolyzed to release its free glutamate content, which is then mixed with salt, vegetable extracts, and other ingredients to produce a dark brown, sticky paste with a strong, salty taste.

About 2 percent of Marmite (1,960 mg/100 g) is made up of free glutamate, giving it an intense umami taste. In addition, as it has significant quantities of vitamin B, including folic acid (B_9) and B_{12}, the paste is a good source of this vital nutrient. After vitamins were first discovered and described scientifically in the early 1900s, Marmite quickly gained popularity. Beriberi, a disease caused by vitamin B deficiency, had been common in Britain during World War I, and Marmite was embraced as a way to prevent it. Its nutritional content was the very justifiable basis for a later marketing campaign that promoted the spread as a source of sufficient vitamin B to keep nerves, brain, and digestion in proper working order.

Marmite is used more or less in the same way as Bovril, often spread thinly on toasted, buttered bread or made into a drink. Whatever the nutritional merits of Marmite, its palatability evokes strong opinions one way or another, and it is definitely an acquired taste. Its fans defend it passionately, while others think that it is disgusting and inedible. The manufacturer capitalized on these reactions a number of years ago, launching a marketing campaign with the slogan "Love It or Hate It."

Another similar product, Vegemite, which has gained the status of a national icon in its home country of Australia, followed a couple of decades later. The name draws attention to its vegetable content, to which autolyzed brewer's yeast is added. In 2009 the parent company, Kraft Foods, attempted to increase the appeal of this traditional product to the younger generation by rebranding it iSnack. The experiment was met with great resistance by the customer base and was quickly abandoned.

There are, naturally, other related pastes that are more locally based, among these Promite and AussieMite in Australia and Cenovis in Switzerland. True connoisseurs are aware of the subtle taste and texture differences that set them apart from each other.

Bovril, Marmite, and Vegemite, protein-rich products with tons of umami.

> **YEAST EXTRACT**
>
> Yeast is an important component in many products, including bread and beer. Yeast cells can be burst open by placing them in a salt solution, causing them to swell up so that the stiff cell walls break apart. Heat is used to break the cell walls down even further. Yeast extract consists of the remaining contents of the yeast cells after the pieces of cell wall are removed.
>
> Yeast extract has a substantial quantity of free glutamate, which is formed when the cells' own enzymes break down the proteins after the cells have ruptured, a process known as autolysis. Additional glutamate can be released by adding enzymes from other sources.

Flakes of nutritional yeast.

NUTRITIONAL YEAST

Another source of umami from yeast is known as nutritional yeast, made up of flakes of dried, inactive yeast. It is produced from ordinary baker's yeast (*Saccharomyces cerevisiae*) that is cultivated on a substrate of molasses from sugarcane or sugar beets. The yeast, along with its enzymes, is deactivated—that is to say, killed—by warming it, followed by washing and drying. The finished product is in the form of a yellowish powder or small flakes.

Nutritional yeast is quite distinct from yeast extract, which is much darker and has a much stronger taste. While nutritional yeast is also a source of glutamate, the total amount is less than that of yeast extract, but it has an abundance of vitamin B and of all the important amino acids.

Many Americans have experienced the amazing effect of nutritional yeast together with salt sprinkled on warm popcorn. Some movie theatres even have it available in shakers as an alternative topping, instead of butter. The umami in the yeast combines well with that in the corn.

Nutritional yeast has a nutty, creamy taste, somewhat like Parmesan cheese, which is why it is often used by vegans as a cheese substitute. It is easy to emulsify nutritional yeast in cold water or other liquids, making it suitable for use in dressings. The taste and consistency of such dressings resembles that of one made with white miso. ⋯> *Dressing with nutritional yeast*

Nutritional yeast flakes have a whole range of uses—for example, on green salads, with gratins, or with shellfish. ⋯> *Eggplant gratinée with garlic, anchovies, and nutritional yeast (page 174).* ⋯> *Oysters au gratin with a crust of nutritional yeast and smoked shrimp head powder (page 175)*

DRESSING WITH NUTRITIONAL YEAST

Nutritional yeast is also jokingly called hippie dust. Appropriately, this recipe comes from an old hippie café in Florida that was popular with vegetarians and vegans.

10 g (¾ Tbsp) nutritional yeast flakes
¾ dL (⅓ c) lemon juice
½ dL (¼ c) tamari soy sauce
½ dL (¼ c) grape seed oil
¼ dL (⅛ c) apple cider vinegar
½ tsp dried thyme or basil
freshly ground black pepper

Mix all of the ingredients together. The herbs can be replaced by flakes of dried sea lettuce. To make a lighter dressing without oil, just mix the yeast flakes with rice vinegar, cooking sake, and small flakes of sea lettuce.

MORE SOURCES OF UMAMI FOR VEGANS

When umami derived from meat, fish, eggs, and dairy products is not an option, one must, like the Buddhist monks who invented the vegetarian temple cuisine *shōjin ryōri*, have recourse to plants and fungi. Obvious sources are dried fungi, which have both glutamate and nucleotides; fermented soybean products, such as soy sauce, miso, and fermented tofu; or fermented wheat protein, such as *seitan*. The easiest way, of course, is to make use of hydrolyzed vegetable protein, yeast extract, and nutritional yeast.

EGGPLANT GRATINÉE WITH GARLIC, ANCHOVIES, AND NUTRITIONAL YEAST

neutral-tasting oil, for deep-frying
eggplants, preferably the firm, long, and thin Japanese variety
cloves garlic, crushed
anchovy paste
nutritional yeast
panko bread crumbs

1. Heat the oil in a large, deep pot over medium-high heat until hot. Preheat the oven broiler.
2. Slice the eggplants in half lengthwise and deep-fry them in the hot oil until they are done through. Allow to cool.
3. Score a diamond pattern into the cut sides of the eggplants and rub them with the crushed garlic. Then spread the anchovy paste on them. Make a half-and-half mixture of nutritional yeast and *panko* bread crumbs and sprinkle that on top.
4. Broil until the surface is golden.

An American classic: Heinz tomato ketchup.

KETCHUP

Ketchup is a purée made from sun-ripened tomatoes. Many households in the Western world have a bottle of ketchup in the kitchen, where it is one of the most frequently used umami enhancers. Although ketchup is often associated with American fast food, such as hamburgers and fries, its roots are actually to be found in the Far East and Indonesia, where it was first fermented as a special type of salty, spicy fish sauce. The origin of the name is unknown, but a variant of it is the name of the original Chinese fish sauce, *koe-chiap*, which refers to the brine in which fish and shellfish were marinated in China.

English sailors brought this Chinese fish sauce back to Europe, where over time the recipe evolved to incorporate fungi, anchovies, tomatoes, vinegar, walnuts, pickled vegetables, and a number of spices. The tomatoes were included to lend umami to what was otherwise a blend of salty, bitter, and sour tastes. Along the way, an increasing amount of sugar was added, rounding out the full complement of all five basic tastes.

Tomato ketchup in its more or less present form dates back to the beginning of the 1800s. The best-known version is Heinz ketchup, which was put on the market in 1876 and is still made according to the same recipe.

OYSTERS AU GRATIN WITH A CRUST OF NUTRITIONAL YEAST AND SMOKED SHRIMP HEAD POWDER

½ dL (¼ c) tomato cut into small cubes
½ dL (¼ c) cucumber cut into small cubes
12 oysters in their shells
½ shallot
1 dL (scant ½ c) dry white vermouth, such as Noilly Prat
2 dL (⅞ c) heavy cream
2–3 Tbsp bread crumbs, preferably *panko*
2–3 Tbsp nutritional yeast
pinch of smoked shrimp head powder
pinch of cayenne pepper

1. Blanch the tomatoes for a minute in boiling water and then skin them. Remove the seeds from both the tomatoes and the cucumber and then cut the pulp into small cubes.
2. Preheat the oven to 225°C (425°F). Open all the oysters and remove them from their shells. Set aside the liquid in the shells and return the oysters to the more curved half-shells. Place on a baking sheet.
3. Chop the shallot finely and place in a saucepan. Add the vermouth, and bring to a boil. Add the cream and reduce the liquid by about half. Add the liquid from the oysters to taste, and stir the tomato and cucumber cubes into the sauce.
4. Mix together the bread crumbs, nutritional yeast, shrimp head powder, and cayenne.
5. Distribute the sauce evenly around and on top of the oysters and sprinkle with the bread crumb mixture.
6. Broil until the crust is golden and the oysters are warm but still raw in the middle.

BAGNA CÀUDA

Anchovies and anchovy paste feature in many Italian dishes, as their significant glutamate and inosinate contents are an excellent source of umami. One has only to think of a Caesar salad to understand why anchovies are often added to sauces and dressings.

A good example of the intense umami that anchovies can impart is the Italian specialty *bagna càuda*, a sauce with a very robust taste from the Piedmont region. It is made from olive oil, garlic, cream or milk, and many anchovy fillets. Typically, the sauce can be used as a type of fondue or as a hot dip for a variety of raw and cooked vegetables, such as bell peppers, fennel, celery, cauliflower, artichokes, and onions. ⋯> *Bagna càuda*

BAGNA CÀUDA

---

10 cloves garlic
4–5 dL (about 2 c) milk
200 g (7 oz) salted anchovy fillets
100 g (3½ oz) butter
2 dL (⅞ c) extra-virgin olive oil
2–3 Tbsp bread crumbs, preferably panko
freshly ground white pepper

1. Peel the garlic. Blanch the cloves for 1 minute in a little milk and water. Rinse under running water. Repeat this process two more times.
2. Bring 1½ dL (⅔ cup) milk (with no water) to a boil in a saucepan, and blanch the garlic one last time. Set aside the milk.
3. Soak the anchovy fillets for 10 minutes in cold water.
4. Mix together in a blender the garlic, anchovies, and butter while adding the olive oil in a steady stream.
5. Mix in the bread crumbs and season with the freshly ground white pepper.
6. Add sufficient warm milk from blanching the garlic to obtain the desired consistency.
7. Serve as a dip for raw vegetables or use as a sauce for fish or poultry.

Worcestershire sauce from Lea & Perrins.

WORCESTERSHIRE SAUCE

Worcestershire sauce is basically a variety of fermented anchovy sauce and, consequently, is related to the classical Roman fish sauce, *garum*. It is a favorite taste enhancer in many Western kitchens. Even famous celebrity chefs will admit that they use this condiment in their creations.

Worcestershire sauce was first made commercially in Worcester, England, in 1837 by two apothecaries, John Wheeley Lea and William Henry Perrins. It caught on quickly and its popularity spread to the colonies as well. Their original recipe was a closely guarded secret for more than 170 years, until 2009. By then, the Heinz food conglomerate had acquired the company, and a former employee at Lea & Perrins claimed to have found some old notes that revealed the hitherto-unknown ingredients in the recipe.

The content label of the modern product lists malt vinegar, spirit vinegar, molasses, sugar, salt, anchovies, tamarind extract, onions, garlic, spice, and flavors. As it turns out, the old notes identified the mysterious 'spice and flavors' as soy sauce, cloves, lemons, pickles, and pepper and indicated the correct proportions of each. What is still missing, however, is a description of how all the ingredients are mixed together. It is said that the workers at the now-defunct Lea & Perrins factory were not allowed to know precisely what was in the sauce, as the ingredients were given to them in code.

There are now several varieties of Worcestershire sauce; for example, there is one without anchovies for vegetarians and vegans.

Worcestershire sauce contains free glutamate (34 mg/100 g) and is commonly used to add umami to marinades, soups, meat dishes, and sauces. It is, of course, an indispensable ingredient in a Bloody Mary or a Bloody Caesar, combining with the tomato juice to turn them into the ultimate umami cocktails.

UMAMI IN A TUBE

In 2010, the upscale British supermarket chain Waitrose launched, with little media fanfare, a new product that was to be the way to the fifth taste and instant umami: Taste No. 5. This trademarked name none too subtly channels that of a well-known perfume. The product is a brown paste sold in an old-fashioned, squeezable tube and described as a flavor bomb. Waitrose's advertisements claim that with Taste No. 5 one can "transform even the blandest meals into something truly extraordinary." The ingredients include Parmesan cheese, anchovies, black olives, tomato purée, porcini mushrooms, and balsamic vinegar, all of which contribute umami and interact synergistically with each other. So in a sense, Taste No. 5 can be said to be Mediterranean umami in a concentrated form. Another Taste No. 5 product has recently been launched based on Asian ingredients, including miso, *yuzu*, shiitake, soy sauce, and *maccha*.

Taste No. 5:
a commercial umami paste.

Twelve easy ways to add umami

Mushroom salt
Cut shiitake or other dark mushrooms into slices and dry them in an oven on low heat. Crush them into a powder and mix it with Maldon sea salt flakes.

Use to season fish, soups, vegetables, and pasta dishes.

Highly concentrated chicken bouillon
1 L (4¼ c) chicken stock reduced to 1 dL (½ c) or less.

Use as an essence in gravies that are a little flat or to add depth to a dressing, or drizzle on pasta or salads.

Marinated mushrooms
Marinate mushrooms in a little soy sauce or garum.

Can be fried or used raw in salads.

Miso paste
Light or dark paste made from fermented soybeans; available where Asian foods are sold.

Adds a nutty, savory taste to dressings, sauces, marinades, and soups (especially those with shellfish); or use it like butter to coat warm vegetables just before serving.

Essence of Worcestershire sauce
Concentrated reduction of the sauce kept at the ready in a small bottle with an eyedropper.

Just add a couple of drops to meat that is being fried or to a sauce or a dressing. Rounds out the taste of a pâté or an egg dish.

Anchovy paste
Available in a squeezable tube to keep in the refrigerator.

For all types of vinaigrettes, dressings, marinades, pesto, and pâtés.

Nutritional yeast

Available in health food stores and many supermarkets.

Use in marinades, sprinkle on vegetarian dishes and grilled vegetables, or mix in with the crumbs used for breading fish. Toast lightly in a skillet with a little olive oil, some ground pure chile powder, and a few bread crumbs and use as a savory topping.

Parmesan cheese crusts

The hard crusts from Parmesan cheese.

Cook with the other ingredients in soups and sauces and remove before serving.

Bits and pieces of air-dried ham

Fat trimmings, bones, and skin.

Throw into the soup or stew pot and remove before serving.

Flatfish dorsal fins

The soft, translucent flesh of the thin muscle of the dorsal fin is cut into small squares and fried lightly.

Use like croutons or bacon bits on a salad; can be marinated with a little lemon juice before serving

Bagna càuda

Made from garlic, olive oil, butter, milk, bread crumbs, and a lot of anchovies.

Use as a dip for raw vegetables or as a sauce for fish or poultry.

Remnants of soft, ripened blue cheese

Place on parchment paper and dry out in the oven or in a food dehydrator.

Grate or crush into a powder and sprinkle on pizza, salad, or pasta, or add to a gravy that tastes a bit flat.

QUINTESSENTIALLY DANISH: BROWN GRAVY, *MEDISTERPØLSE*, AND BEEF PATTIES

Brown gravy, *medisterpølse* (pork sausage), and beef patties are among the most loved of the dishes that have been staples of ordinary Danish cuisine for centuries. While these are truly prime examples of 'grandmother's cooking,' there is something more to them. Actually, they are fine illustrations of the intuitive search for umami that has been going on in Western kitchens for hundreds of years. And in light of our recent knowledge about the fifth taste, they can be made even better by tweaking the combination of ingredients that go into them.

Both the taste and the aroma of an old-fashioned brown gravy are a heavenly experience, the very essence of umami. When made properly, it is based on a stock made from meat and bones in combination with vegetables that have simmered for a long time to draw out every bit of free glutamate and inosinate in them. When possible, use water in which potatoes have been boiled as well, because it is full of the free glutamate released from them during cooking. To make it even more savory, one can slip in an anchovy right at the start. The fish totally melts away into the sauce, adding an extra, intense infusion of umami. Its own taste disappears so completely that no one will ever guess that 'grandmother's brown gravy' has been doctored in this untraditional way.

Conversely, a really unappetizing brown gravy is worse than no sauce at all, and just about everybody can tell when the taste is not what was expected. Something is missing, and most often this comes down to salt and umami. Several measures can be undertaken to rescue it. Adding stock as above will automatically make it taste saltier. The gravy will be improved even more if fungi, which can contribute guanylate and interact synergistically, are added to the mix. If it simply is a bit flat, an easy way to pump up the umami content is to add a dash of Worcestershire sauce or a small lump of mature blue cheese.

Medisterpølse is a traditional Danish aromatically spiced dinner sausage made from ground pork and striped pork fat. The secret behind its rich umami content is mixing the meat with a stock made from pork bones before stuffing it into the sausage casings. It is usually eaten with plain boiled potatoes and a thick brown gravy, which elevates the taste of the dish to a whole other level. As mentioned, the cooking water from the potatoes should be set aside for the gravy. ⇢ *Old-fashioned Danish medisterpølse*

OLD-FASHIONED DANISH *MEDISTERPØLSE*

Stock

½ kg (1 lb) pork bones
bouquet garni with celery

salt and freshly ground black pepper
a few whole cloves (optional)

Stuffing

1 kg (2¼ lb) pork, not too lean
250 g (9 oz) striped pork fat
4 shallots, peeled

salt and freshly ground black pepper
ground allspice
sausage casings

1. Place the stock ingredients in a pot with a tight-fitting lid, add just enough water to cover, and simmer for 1 hour. Strain the liquid and reserve for use later.
2. Grind the pork, pork fat, and shallots coarsely in a meat grinder.
3. Stir in enough of the stock to give the meat the consistently of porridge, and mix well. Season with a little salt, pepper, and allspice.
4. Using a sausage horn mounted on a mixer or grinder, stuff the sausage casings. Be careful not to overfill the casings, or they will burst when the sausages are cooked.
5. Divide the sausage into appropriate lengths and tie a knot in each end.
6. Because these sausages are not cooked or smoked, they should be frozen if not intended for use within a day or so.
7. *To serve:* Boil the sausages in a very small amount of water until they are cooked through (you will have to cut into one to check). Reserve this water, together with water from boiling potatoes, for making brown gravy. Fry the cooked sausages in butter until they are nicely browned on all sides. Serve with brown gravy and boiled or steamed potatoes, with some cucumber salad or pickled beets on the side.

The third in this trio of everyday dishes consists of beef patties with onions, potatoes, and brown gravy. In terms of old-fashioned Danish cuisine, it represents the very epitome of food with umami. When Danes are asked by foreigners to identify food that is quintessentially representative of their country, they often answer *smørrebrød* (smorgasbord or open-face sandwiches) and beef patties with onions. The patties are made with ground beef, which should be neither too fatty nor too lean. A fat content of about 8 percent is best. Ideally, the meat should have been aged and be from older animals, as this results in the most umami.

Usually the patties are served with onions fried until they are translucent and a brown gravy made with the meat drippings from the frying pan. Here we have updated the recipe to increase its umami content by seasoning the gravy with Worcestershire sauce, HP sauce, or soy sauce. And, whether one likes it or not, custom dictates that the gravy must be a deep brown color, so it might be necessary to add a little caramel food coloring to achieve the desired result. Boiled or mashed potatoes and a fried or poached egg, which also contribute umami, are served on the side. The dish is rounded out with condiments that are sweet and sour, such as pickled beets, cucumber pickles, sweet and sour red cabbage, and wild cranberry or red currant jelly. ⋯> *Beef patties, Danish style*

Beef patties, Danish style, with onions, mashed potatoes, marinated beets, poached egg, and brown gravy.

SLOW COOKING: THE SECRET OF MORE UMAMI
Gently stewing or simmering meat and vegetables over a long period of time releases more of their free glutamate and nucleotides, which intensifies umami. A dish of oxtails simmered for hours or an osso buco made with beef shanks, vegetables, and tomatoes is delightfully savory and satisfying.

The difference in taste between a well-cooked meat stew and a quickly sautéed piece of beef is enormous. Sautéing will infuse the dish with delicious-tasting Maillard compounds as the result of browning, but stewing draws out more umami. Other traditional simmered dishes that are also rich in umami include cabbage rolls and cassoulet. ⋯> *Cassoulet (page 186)*

The French word *ragoûter*, meaning 'to revive the taste,' precisely describes the effect of umami. It has been absorbed into English as ragout, another name for a stew that is often applied to a variety of delicious dishes made with meat, poultry, or game. *Estofado* is a traditional Spanish casserole in which meat is stewed with tomatoes and potatoes. ⋯> *Beef estofado (page 188)*

BEEF PATTIES, DANISH STYLE

large baking potatoes
salt
mustard seeds
beets
apple cider vinegar
canola oil
freshly ground white pepper
Worcestershire sauce
1 egg per serving
onions
ground beef with about 8% fat content, about 200 g per serving
butter
all-purpose flour
Worcestershire sauce, soy sauce, HP sauce, or ketchup
caramel food color (optional)

Serves 4

1. Wash the potatoes and boil them whole with 4 g (⅘ tsp) of salt per liter (quart) of water. Set aside. Remember to set the cooking water aside as well.
2. Soak the mustard seeds in a little lukewarm water to soften, and then crush them lightly.
3. Cook the beets in lightly salted boiling water for about 45 minutes, until they are tender. Discard the water and place the beets under cold running water to loosen the skin. Peel or slip off the skin. Grate the beets or cut them into strips, then marinate them with the mustard seeds, apple cider vinegar, and canola oil, as well as a little salt and white pepper.
4. Pour the Worcestershire sauce into a small pot, reduce by about half, and allow to cool.
5. Place the eggs in egg poachers and add a little salt, white pepper, and a drop of the Worcestershire reduction.
6. Poach the eggs gently, 4–5 minutes. The yolks should be runny.
7. Peel the onions and cut them into chunks, like orange segments.
8. Shape the ground beef into patties, fry them in butter, and season with a little Worcestershire sauce reduction, salt, and white pepper. When you turn the patties, add the onion segments to the pan.
9. Remove the patties when they are medium-rare. Leave the onions to fry until they are golden and soft. Remove the onions from the pan and place most of the onions on top of the patties. Reserve the rest for serving.
10. Sprinkle a little flour in the pan, toast it lightly, add two or three ladlefuls of the potato water, and whisk thoroughly. When the mixture has cooked through and the floury taste has disappeared, add more potato water to attain the desired consistency for the gravy.

11. Season the gravy with Worcestershire sauce, soy sauce, HP sauce, or ketchup, as well as salt and white pepper. Add a little caramel food coloring to give color—not necessary, but not harmful either!
12. Cut the cooked potatoes in half, hollow them out, put the potato flesh through a ricer or mash them with a fork.
13. Arrange the beef patties on plates with the beets, remaining onions, poached eggs, mashed potatoes, and the brown gravy.

Chicken Marengo.

Adding shellfish to a meat stew that has tomatoes introduces synergistic umami from the abundance of nucleotides in these ingredients. An excellent example of this type of dish is chicken Marengo. Legend has it that it was created following the Battle of Marengo in Italy in June 1800, when the French led by Napoleon were fighting the Austrians. In order to feed the troops after their victory, Napoleon's chef, Dunand, sent his men out to forage in the surrounding countryside. They returned with tomatoes, eggs, chicken, and crayfish. The chef created a delicious stew, which so delighted Napoleon that he allowed it to be named after his triumph at Marengo and had it served to him after every battle. The combination of chicken, eggs, tomatoes, and crayfish impart a very intense umami taste. ⋯> *Chicken Marengo*

A fricassee is a stew made with light meat—for example, veal, poultry, or lamb—together with a variety of vegetables and, possibly, mushrooms. A classic dish is chicken fricassee, which is rich in umami from the glutamate in the chicken and vegetables and the guanylate in the mushrooms. Served in small pastry shells, chicken fricassee with asparagus is a delicious treat.

CHICKEN MARENGO

1 free-range chicken, about 1.5 kg (3 lb)
olive oil
12 fresh crayfish
20 pearl onions, peeled
2 cloves garlic, crushed
2 dL (⅘ c) dry white wine
1 can (2 c) peeled, chopped tomatoes
2–3 dL (⅘–1⅕ c) chicken bouillon or dashi
5 sprigs fresh thyme
country bread
salt and freshly ground black pepper
4 free-range eggs
200 g (7 oz) chanterelles or other tasty mushrooms
1 bunch fresh parsley

Serves 4

1. Split the chicken open and discard the backbone. Cut the rest into pieces, complete with the carcass and the bones, which add taste substances.
2. Brown the pieces thoroughly in olive oil in a large pot and set aside.
3. If necessary, add a little more olive oil to sauté the crayfish for a few minutes. Take out the crayfish, remove their heads, and set them aside. Put the heads back in the pot together with the onions and garlic. Cook until the onions are light brown, then add the white wine and reduce the liquid a little.
4. Return the chicken pieces to the pot. Add the crushed tomatoes, chicken bouillon, and thyme, cover with a lid, and allow to simmer over low heat until the chicken is cooked through but still juicy. Remove the carcass and bones and keep the chicken warm until serving.
5. Peel the crayfish. Cut the crust off the bread, and cut it into cubes. Toast first the crayfish and then the bread cubes lightly in olive oil in a skillet.
6. Fish out the crayfish heads from the pot with the chicken and discard; season the sauce with salt and pepper.
7. Crack the eggs into a bowl filled with cold water and vinegar so that they tighten up. Warm some olive oil in a small pan. Remove the eggs carefully and pat them dry on a paper towel. Fry them in the oil until they are golden but still runny inside.
8. Place the eggs on a paper towel to drain the excess oil and season them with sea salt and freshly ground pepper.
9. Toast the chanterelles in a dry skillet. Chop the parsley.
10. Serve the chicken in the tomato sauce. Distribute the bread cubes over the chicken. Add the fried eggs, sprinkle with the parsley, and top with the crayfish.

CASSOULET

Serves 4

300 g (10 ½ oz) dried white beans
250 g (9 oz) carrots
4 onions
8 cloves garlic
500 g (17½ oz) ripe tomatoes
200 g (7 oz) pork crackling
olive oil
fresh thyme leaves
4–6 confit duck legs (can be purchased ready-made)
500 g (17½ oz) coarse sausages, preferably small ones with a lot of meat
600 g (21 oz) free-range pork shoulder
300 g (10½ oz) smoked bacon, pieces halved
bay leaves
whole cloves
salt and freshly ground black pepper
1½ L (6⅓ c) soup stock or water

1. Soak the beans for about 12 hours, changing the water once. Drain the beans.
2. Peel the carrots and cut into pieces. Chop the onions and mince the garlic finely.
3. Blanch the tomatoes in boiling water for 1 minute, then remove their skin and chop them.
4. Chop the crackling into small pieces and sauté it in olive oil in a skillet over medium-low heat with half of the onions and garlic, without letting it brown.
5. Add the drained beans and a little thyme and allow it all to simmer uncovered for about 2 hours.
6. In another skillet, sauté briefly the duck legs in their own fat together with the sausage, which can be cut into large pieces. Remove from the pan.
7. Cut the pork shoulder into chunks and sauté in the same pan together with the bacon pieces.
8. Add the remaining onions and garlic, carrots, tomatoes, thyme, bay leaves, a few cloves, salt and pepper, and soup or water.
9. Allow to simmer for about 2 hours.
10. Preheat the oven to 160°C (320°F). Cut the bacon into smaller pieces and distribute all of the meat pieces and other ingredients evenly in a clay baker and cover with the soup stock or water.
11. Bake, adding a little more liquid if necessary during baking, until it has a little bit of a crust and the beans have absorbed all the liquid.
12. Serve directly from the oven, accompanied by slices of coarse country bread.

▸ Cassoulet.

Making the most of umami

BEEF ESTOFADO

Serves 4

For the red wine marinade

500–600 g (17½–21 oz) root vegetables, such as celery, carrots, onions, and leeks
olive oil
1 bottle (750 ml) dry red wine
1 dL (⅖ c) red wine vinegar
fresh parsley stalks, sprigs of fresh thyme, bay leaves, black peppercorns, and peeled garlic cloves
300 g (10½ oz) bacon
1½ kg (3⅓ lb) beef chuck roast, bone-in

For braising and serving

root vegetables, such as celery, carrots, and celeriac
generous dash of brandy
a little more red wine, if needed
crushed tomatoes
mashed potatoes

1. Cut the first lot of root vegetables into large pieces and toast them in a pot in a little olive oil. Add the red wine, red wine vinegar, and the herbs and spices. Allow to simmer, covered, for 15–20 minutes; then cool.
2. Divide the bacon into two pieces, cut the meat into 8 pieces. Immerse the pieces completely in the cooled marinade and let sit in a cool place for 24 hours, turning them once.
3. Preheat the oven to 130°C (260°F). Allow the bacon and meat to drain thoroughly and then brown it well in a large skillet. Transfer to an ovenproof baking dish. Cut the second batch of root vegetables into large pieces and add to the dish. Pour in the marinade and add a generous dash of the distilled spirits and possibly a little more red wine. As the meat should be covered with liquid, it might be necessary to add a little beef bouillon or even a bit of water.
4. Braise in the oven for about 3 hours until the meat come off the bone easily.
5. Strain the liquid into a pot and skim off the fat carefully with a spoon. Reduce the liquid a little over high heat. The gravy can be thickened *en roux* using the fat, or by blending some of the vegetables and adding them to the gravy. What the gravy loses in appearance will be compensated for by what it gains by way of taste.
6. *To serve:* Prepare a large portion of crushed tomatoes as follows: Blanch the tomatoes in boiling water for 1 minute and peel them. Remove the seeds and chop the pulp into small cubes. Mix the tomatoes into the *estofado* and serve it with mashed potatoes, preferably mixed with puréed root vegetables.

▸ Beef *estofado*.

RATATOUILLE AND *BRANDADE*

The traditional Provençal ratatouille made with a variety of vegetables, typically eggplants, tomatoes, onions, and celery simmered in olive oil, can be a little on the heavy side. There is a Sicilian variation, however, which is lighter in color and less filling. ⇢ *Sicilian ratatouille*

Sicilian ratatouille can be served as an accompaniment to a *brandade*, which is made with salt cod. ⇢ *Brandade with air-dried ham and green peas*. The combination of these two dishes hits all the right notes, with umami from tomatoes, dried ham, and green peas.

SICILIAN RATATOUILLE

Serves 4

400 g (14 oz) eggplants
2–4 fresh artichokes
200 g (7 oz) zucchini
5 stalks celery
salt
olive oil
200 g (7 oz) shallots, finely chopped
2 cloves garlic, crushed

400 g (14 oz) very ripe tomatoes
2 Tbsp tomato purée
2 Tbsp drained capers
chopped fresh oregano
white wine vinegar
sugar
freshly ground black pepper
2 tsp shelled pistachios

1. Cut the eggplants, artichokes, zucchini, and celery into cubes. If fresh artichokes are not available, oil-preserved artichokes in a jar can be substituted.
2. Salt the eggplant cubes lightly and place them in a colander to drain for about 20 minutes. Rinse and pat them dry.
3. Sauté the eggplant cubes in a pot with olive oil until browned, and then remove them. Next sauté briefly the shallots, garlic, zucchini, celery, and artichoke cubes in the pot. Add the tomatoes and the tomato purée to the pot. Cover and allow to simmer for about 10 minutes.
4. Add the eggplant cubes, capers, and oregano to the pot. Season with a little white wine vinegar, sugar, salt, and pepper.
5. Toast the pistachios in a dry skillet and sprinkle over the ratatouille. It tastes even better if it has been allowed to sit for a day before eating.

BRANDADE WITH AIR-DRIED HAM AND GREEN PEAS
---

300 g (10½ oz) salt cod
olive oil
3 cloves garlic
10 white peppercorns
2 bay leaves
1 sprig fresh thyme

4 starchy baking potatoes, peeled and cut into cubes
1 onion, finely chopped
100 g (4 oz) air-dried ham, cut into cubes
green peas or green asparagus

Serves 4

1. Soak the salt cod in water for 24 hours, changing the water a few times.
2. Taste the fish. It should be neither too salty nor too waterlogged.
3. Place the salt cod in a stainless steel bowl with the skin side up. Heat some olive oil with the garlic, peppercorns, bay leaves, and thyme to 150°C (300°F) and pour over the salt cod. Allow to cool to room temperature.
4. Remove the skin and bones from the fish.
5. Pour a little olive oil into a pot. Add the potatoes and onion. Allow to fry gently over low heat for a few minutes without browning. Add 1 cup water and simmer with the cover on until the potatoes are soft.
6. Squash the potatoes with a fork to make a coarse mash. Stir the salt cod and cubes of air-dried ham into the mash. Season with olive oil, salt, and pepper. Add raw green peas or finely cut asparagus at the very end.
7. Serve the *brandade* in scoops. It goes well with ratatouille and can be eaten warm, lukewarm, or cold as a dish on its own with toasted bread and olive oil.

THIS IS WHY FAST FOOD TASTES SO GOOD

Having just described slow cooking as a sure way to unlock maximum taste, we might quite naturally ask why the stereotypical fast foods can still taste good, even when they are made from inferior ingredients or using industrial techniques that treat them in a rather brutal manner. Those that have little salt, or less fat, or substitute lean meat may have fewer calories, but they lose a great deal of taste, which is in the fats or enhanced by the addition of salt. So something is needed to fill the gaps, and the fast food industry often depends on umami to solve many of these problems. Much fast food is rendered edible, and may even be delicious, simply because it combines ingredients that maximize savory tastes. This is true for a hamburger with slices of tomato, ketchup, mayonnaise, and cheddar cheese; deep-fried potatoes with ketchup, spaghetti with tomato sauce; pizza with tomatoes and Parmesan cheese; or noodles with soy sauce. ⇢ *Three-day pizza with umami—not really a 'fast food' (page 192)*

UMAMI BURGER
An ordinary hamburger is the stereotypical example of fast food—a plain, soft bun and a dry patty made from low-quality meat. It is edible only because the ketchup, cheddar cheese (sometimes), tomato slices, bacon strips, and mayonnaise that accompany it are full of glutamate and impart umami. An American burger restaurant on the West Coast and in New York, which calls itself Umami Burger, has achieved nearly cult status by going all the way and actively embracing the concept of the fifth taste. With its slogan of "It's all about umami," the restaurant serves hamburgers with tomatoes, konbu, anchovies, shiitake mushrooms, truffles, Parmesan cheese, and soy sauce. A veritable umami fix.

THREE-DAY PIZZA WITH UMAMI—NOT REALLY A 'FAST FOOD'

Serves 4

Pizza dough

Day 1

2 dL (⅞ c) water
400 g (1¾ c) high-gluten flour, such as durum
a knifepoint of active dry yeast

1. Mix the ingredients together into a *biga* and leave covered in a cool spot to rise until the next day.

Day 2

Biga from Day 1
3 dL (1¼ c) water
2 g (⅓ tsp) active dry yeast
3 Tbsp olive oil
350 g (1½ c) fine durum flour
150 g (⅔ c) coarse whole wheat flour
1½ tsp salt

2. Mix the *biga*, water, yeast, salt, and olive oil thoroughly so that the *biga* is completely blended in.
3. Add both flours a little at a time and knead thoroughly to make a firm dough. The amount of flour needed may vary.
4. Set aside covered in a cool place to rise until the next day.

Day 3

5. Punch down the dough and cut it into quarters. Put each piece in a bowl and allow the dough to rise until double its size.

Basic tomato sauce

1 large onion
olive oil
2 cans (3–4 c) peeled tomatoes
3 Tbsp tomato paste
1 dL (⅖ c) dry white wine
fresh oregano leaves
sugar
salt and freshly ground black pepper

6. Slice the onions very thinly and cook them in some olive oil in a pot over medium-high heat until they are translucent.
7. Add the tomatoes, tomato paste, wine, and oregano and allow to simmer for about 20 minutes uncovered. The sauce can be puréed if a finer consistency is desired.
8. Season with sugar, salt, and pepper. Set aside to cool.

Pizzas with an abundance of umami from tomato sauce, air-dried ham, anchovies, and Parmesan cheese.

Baking the pizzas

9. Place a pizza stone in the oven and preheat the oven to the highest temperature, preferably above 300°C (550°F). Roll out the pizza dough pieces and spread a layer of the tomato sauce on top. Cover with the toppings.
10. Bake the pizzas on the preheated pizza stones (or on a pizza pan).

Some suggested toppings for the pizzas are: crumbled Gorgonzola, anchovies in olive oil, pieces of fresh asparagus, dark mushrooms according to season, capers, Parmigiano-Reggiano, pancetta, air-dried ham, and truffles (grated on top after baking).

There are pizzas and, then again, there are pizzas. If, however, one has the great good fortune to be on hand when everything comes together flawlessly, pizza rises above the ordinary. If the dough, the sauce, and the toppings all complement each other perfectly and the pizza is baked in a wood-fired stone oven, which produces a crisp crust with air bubbles and black smudges, and it has a true smoky taste, one has had the chance to experience the essence of deliciousness and simplicity.

GREEN SALADS AND RAW VEGETABLES

A green salad is an example of food that is fairly uninteresting on its own. It is lacking something, and that something is often umami. The classical example is a Caesar salad made with leaves of plain green romaine. Its taste is due almost entirely to the dressing and toppings that accompany the lettuce. These can be Parmesan cheese, cooked eggs, anchovies, ripe tomatoes, crisp bacon bits, or Worcestershire sauce, all ingredients that impart umami.

Many people choose not to eat vegetables because they consider them to have little taste or simply because they do not know how to prepare them so that they are palatable. Just as with a plain green salad, there is probably little that is less appealing than a collection of vegetables to be eaten raw—wholesome, but bland. Once again, one can take a lesson from the Buddhist monks and their use of dashi to make delicious, strictly vegetarian temple meals. Finely sliced vegetables can be simmered in dashi, infusing them with an immediate jolt of umami, which may even interact synergistically with the small amounts of glutamate and ribonucleotides in them to enhance their taste.

▸ Green salad with ingredients that add umami.

A modern version of roasted thrush with an umami-rich gravy, here made with quails.

UMAMI IN DISHES MADE WITH SMALL FOWL

Many cooks know about the tried-and-true trick of adding some soft, ripened blue cheese to a gravy for a dish made with small game birds. For example, ripe, creamy Roquefort or Danish blue cheese is full of free glutamate that will enhance the savory taste of the gravy. At the same time, the cheese adds a little touch of acidity that can be balanced by some fruit jelly, typically from red currants. This same combination is often found in old Nordic recipes for small game birds. A variety of small birds can also be prepared as a pâté. ⋯> *Quail pâté*

QUAIL PÂTÉ

Serves 4

250 g (9 oz) meat from small fowl, such as quails, pigeons, and partridges
100 g (3½ oz) pork lard
250 g (9 oz) calf liver
100 g (3½ oz) mushrooms
2 shallots, finely chopped

5–10 whole madagascar green peppercorns
fresh thyme leaves, to taste
½ dL (¼ c) port
1 egg
salt and freshly ground white pepper

1. Preheat the oven to 155°C (310°F). Grind the poultry meat, lard, liver, and mushrooms in a meat grinder.
2. Mix together with the shallots, peppercorns, thyme, and port. Bind with an egg and season to taste with salt and pepper. Pack into a loaf pan.
3. Place the loaf pan in a larger baking pan and add hot water to the larger pan to a depth of 2.5 cm (1 in). Bake the pâté in the water bath for 30–40 minutes.

COOKED POTATOES: NOTHING COULD BE SIMPLER

Even though potatoes may seem a little uninspiring, they are probably the most striking illustration of the meaning of umami in European peasant cuisine. The taste of cooked potatoes owes much to the umami substances glutamate and guanylate, which are released during cooking. In fact, the free glutamate content of a potato doubles when it is cooked. A cooked potato has practically no aroma, and it is not possible to simulate its taste by combining sour, sweet, salty, and bitter tastes.

Cooked potatoes are often found in dishes that are rich in umami, such as the braised potatoes in cassoulet or in red flannel hash. A dash of Worcestershire sauce or soy sauce adds some extra umami to the potatoes. The water from boiled potatoes can be recycled to advantage for making dashi or gravy.

RICE AND SAKE

How does rice end up with any umami taste when it has none in either its raw state or its cooked state? The addition of MSG or salt does not improve things noticeably, which is natural, given that neither the MSG nor the rice has any taste in and of itself. On the contrary, if one adds soy sauce, which contains a whole series of amino acids but no inosinate or guanylate, the rice becomes slightly tastier. It really starts to improve when one incorporates *katsuobushi*, which is an ingredient in some recipes for sushi rice, or another ingredient that also has a nucleotide to interact synergistically. This we know from those rice dishes that develop a delicious taste when the rice is combined with vegetables, omelette, fried or cooked meat, chicken bouillon, fish, or shellfish. ⋯> *Risotto*

RISOTTO

To make a good risotto, you need three things—the right kind of rice, a lot of enthusiasm, and the ability to pay attention. It is vital to use a special type of round short- or medium-grain rice with a high starch content; Arborio, Baldo, Carnaroli, and Vialone Nano are all very suitable. Just as is the case for dashi, the exact way a risotto turns out is very likely to vary from one kitchen and one cook to another. The central, critical feature, however, is the texture—it must be creamy and so soft that it is almost runny, while at the same time, each grain of rice should retain just a tiny bit of crunchiness.

▸ Risotto with black truffles.

Leftover risotto can easily be used to make little rice fritters. Spread the risotto about 2 cm (¾ in) thick on a piece of plastic wrap on a baking sheet and refrigerate. Cut into appropriate-size pieces, flour them on all sides, and fry them in butter until golden and crisp. Eat as a snack or as part of a meal.

---------- ASPARAGUS RISOTTO ----------

Serves 4

½ kg (17½ oz) white asparagus
2 Tbsp olive oil
2 shallots
250 g (1 c) risotto rice
1 L (4¼ c) chicken bouillon

50 g (3½ Tbsp) butter
75–100 g (⅓–⅖ c) finely grated Parmigiano-Reggiano
salt and freshly ground white pepper

Note: When making risotto, ensure that the liquid to be added is warm.

1. Peel the white asparagus and cut the spears into small pieces. Cook them in ½ L (2⅛ c) lightly salted boiling water for about 2 minutes and then remove from the pot. Reserve the water for the risotto.
2. Heat the olive oil in a heavy pot over low heat. Chop the shallots finely and cook them in the oil until they are translucent. Add the rice and allow it to cook with the shallots for a few minutes.
3. Gradually add the chicken bouillon, a little at a time, and stir carefully until the rice has absorbed the liquid. Alternate between adding the bouillon and the asparagus water until the rice is cooked but has just a little crunch in the middle of each grain.
4. Mix in the asparagus pieces, possibly with a little more bouillon. Remove the pot from the stove and stir in the butter, a little at a time, and the Parmesan cheese. Season to taste with salt and pepper.
5. Cover the pot and allow the risotto to stand for a few minutes before serving.

RISOTTO WITH BLACK TRUFFLES

This is made using the same techniques as the asparagus risotto. Instead of the asparagus water, use a stronger bouillon, for example, from beef, or reduce chicken stock to make it more robust.

2 Tbsp olive oil
2 shallots
250 g (1 c) risotto rice
2 dL (⅞ c) dry white wine
1½ L (6⅓ c) strong bouillon
1 L (4¼ c) chicken bouillon

50 g (3½ Tbsp) butter
75–100 g (⅓–⅖ c) finely grated Parmigiano-Reggiano
salt and freshly ground white pepper
black truffles, according to taste and affordability

Serves 4

Follow the method for the asparagus risotto, except add the white wine to the rice after cooking it with the shallots. Allow the rice to absorb the white wine before adding the bouillon.

At the end, cut the truffles into thin slices. Mix half of the slices into the risotto when it is completely finished and sprinkle the rest on the plates just before serving.

Sake, an alcoholic beverage produced from polished rice, is popularly thought of as a rice wine, even though the fermentation process involved is actually closer to that used for brewing beer.

Quite surprisingly given its rice base, sake has a reasonable umami content. When rice grains are polished, they usually lose at least half of their mass, as well as most of their proteins and fats, leaving behind mostly starch. The polished rice is cooked and fermented with the help of a fermentation medium (*kōji*), which contains enzymes that can break down the starch to sugar and the proteins to free amino acids. A yeast culture then converts the sugar to alcohol. The free amino acids found in the finished sake are derived from the small amounts of protein that remain in the polished rice, as well as those in the *kōji* and the yeast.

The umami content of sake depends on the amount of free glutamate in a particular type and brand. This information is listed on the label of some high-quality sake. Umami is more pronounced in sake that is less dry and slightly sweet. Many Japanese feel strongly, and perhaps quite justifiably, that rice wine pairs perfectly with a meal, as its umami serves to reinforce the delicious taste of the food. It is possible, however, that some of the subtle umami taste so highly prized by sake connoisseurs is not due to the free glutamate in the liquor. Instead, it may be due to the presence of succinic acid, which has a similar effect by introducing a combination of salty, bitter, and sour tastes.

The lees from sake fermentation, called sake *kasu*, are made up of leftover starch and sugar, as well as spent yeast cells, which contain very large quantities of amino acids and glutamate. Consequently, the lees can be recycled to advantage to impart umami; for example, as a marinade for fish or vegetables.

The practice of using sake *kasu*, together with mirin, salt, and sugar, to make a marinade goes back at least twelve centuries in Japan. Traditionally, the foods could be left to marinate for a very long time, in some cases, a period of years. The process helps to preserve the raw ingredients and improves both their keeping qualities and their nutritional value. Originally, sake *kasu* was used to marinate melon, eggplants, and cucumbers; later on, carrots and ginger were also preserved this way.

The combination of sake and mirin is also used to improve the taste of bland vegetables. When marinated, they become slightly sweet and can be eaten cold or as a condiment.

When fish are marinated in sake *kasu*, no sugar is added to the marinade. Lemon juice, sake, ginger, and possibly a little oil with a mild taste can be incorporated in the marinade. Depending on the thickness of the fish, it should marinate from a half an hour to an hour. To serve, remove the fish from the marinade, but do not rinse it or scrape it clean. If the fish is sashimi grade, it can be eaten as is; otherwise, fry it gently over low heat.

BEER

The brewing process used to make beer helps to release small quantities of glutamate (2–4 mg/100 g) from the proteins in the grains that make up the malt. Additional free glutamate comes from the yeast. Unfiltered wheat beer with residual yeast has more glutamate than a pilsner-style beer. For this reason, wheat beer is a good source of umami when marinating or cooking meat and vegetables. ⋯> *Oxtails braised in wheat beer*

OXTAILS BRAISED IN WHEAT BEER

2 kg (4⅖ lb) middle pieces of oxtail, cut into segments
salt and freshly ground black pepper
1 or 2 fresh chile peppers
assorted vegetables, such as celery, carrots, onions, and leeks
200 g (7 oz) shallots
2 bottles (355 mL/12 oz each) unfiltered wheat beer
1 dL (⅖ c) red wine vinegar
100 g (7 Tbsp) tomato purée
1–2 L (4¼–8½ c) beef bouillon or water
2 Tbsp unsweetened cacao powder
a little all-purpose flour
bouquet garni of fresh lovage, thyme, savory, parsley stalks, and dried bay leaves

Serves 4

1. Rinse the oxtails thoroughly, dry them, and season generously with salt and pepper.
2. Cut open the chile peppers lengthwise and remove the seeds.
3. Cut the vegetables into chunks and chop the shallots finely.
4. Brown the oxtails in a large pot with a heavy base, then add the vegetables, bay leaves, shallots, and chile peppers. Sauté them until they have a little color.
5. Pour in the beer and the vinegar and cook until the liquid has almost disappeared.
6. Spread the tomato purée on the tails and add the bouillon until the pieces are just covered.
7. Cover the pot and allow the oxtails to cook over very low heat on the stove top (or in a low oven) for about 3 hours. If necessary, add a little extra liquid from time to time to ensure that the oxtails are just covered.
8. Sprinkle on the cacao powder and allow the oxtails to finish braising.
9. Remove the oxtail pieces from the pot, pass the cooking liquid through a sieve, and skim off any foam or other bits. Reduce the liquid and season to taste.
10. To thicken the gravy, sprinkle a little flour on top, allow the surface fat to absorb it, whisk thoroughly, and cook through until the taste of the flour disappears.
11. *To serve:* Remove the meat from the bones. Serve with the gravy, a good mash, roasted root vegetables, mushrooms, and pickled green beans.

UMAMI IN SWEETS: SOMETHING A LITTLE UNEXPECTED

In Western cuisines, most desserts are made with a considerable quantity of sugar. The sugar adds sweetness and ensures that desserts made with acidic fruits have a well-balanced taste. It is noteworthy that Japanese cuisine has no tradition of adding sugar to desserts. The sweet tastes in a cake or a candy come from the raw ingredients themselves, such as small red azuki beans.

Here, again, umami comes into the picture. It can interact with sweet tastes, making them more pronounced, with the result that even desserts with only a little sugar seem to be sweeter. One can take advantage of this relationship by using raw ingredients with umami and little taste of their own to reduce the amount of sugar in a traditional dessert, such as a sorbet. Experiments have shown that one can substitute up to 25 percent of the sugar in a sorbet with a suitable quantity of the juice from sun-ripened tomatoes. ⇢ *Umami sorbet with maccha and tomato*

UMAMI SORBET WITH *MACCHA* AND TOMATO

Serves 4

15 g (½ oz) maccha
8 dL (3⅓ c) warm water (75–80°C or 165–175°F)
1 leaf gelatin (¾ tsp powdered gelatin)
140 g (⅗ c) light cane sugar

70 g (⅓ c) corn syrup
2 dL (⅘ c) tomato juice from the pulp of very ripe tomatoes
seeds from ½ vanilla bean
a little freshly squeezed lemon juice

1. Pass the *maccha* powder through a sieve into a deep, warmed bowl, add the warm water, and whisk for 1–2 minutes, until the tea foams.
2. Soften the gelatin in a small amount of cold water.
3. Pour the warm tea into a blender, add the cane sugar, corn syrup, gelatin, tomato juice, and seeds from the vanilla bean. Blend until smooth.
4. Season with a little lemon juice, until the sweet and sour balance is as desired.
5. Churn the mixture in an ice cream maker according to the manufacturer's instructions. Or freeze in it a shallow baking pan and break it up and blend it again just before serving. The sorbet can be served with small grape tomatoes that have been pickled in a good, strong honey with some lemon peel.

Umami sorbet with *maccha* and tomato.

Chocolate and chocolate creams are often found in sweet desserts. Adding components with umami, such as blue cheese and nutritional yeast, can enhance the sweetness in addition to imparting the dessert with an element of surprise. ⋯> *White chocolate cream, black sesame seeds, Roquefort, and brioche with nutritional yeast*

MIRIN IS A SWEET RICE WINE WITH UMAMI

Mirin is a sweet liquor made from rice, with an alcohol content of about 14 percent and a significant amount of umami. It is made from cooked rice that is seeded with the fungus *Aspergillus oryzae*, which is also used for the production of Japanese *shōyu* and miso. *Shōchū*, distilled rice wine brandy, is added to the rice. The enzymes in the fungus break down the starch in the rice to sugar and its proteins to free amino acids, including glutamate, which imparts umami. Mirin is not intended to be drunk. A mixture of mirin, soy sauce, and sake is called *tsuyu*, a sauce with a great quantity of umami that is used for dipping tempura and noodles. The well-known, sticky teriyaki sauce is made by cooking this mixture with sugar.

WHITE CHOCOLATE CREAM, BLACK SESAME SEEDS, ROQUEFORT, AND BRIOCHE WITH NUTRITIONAL YEAST

Brioche

It is difficult to prepare brioche in small quantities. Hence the present recipe will yield more than is needed. The baked brioche can easily be stored in the freezer.

½ kg (2⅕ c) all-purpose flour
90 g (⅖ c) sugar
40 g (2¾ Tbsp) nutritional yeast
8 g (1⅔ tsp) salt
5 g (1 tsp) active dry yeast

65 g (¼ c) water
5 eggs
250 g (8¾ oz) butter, at room temperature, cut into small pieces

Serves 4

1. Place all of the ingredients except for the butter in the bowl of an electric mixer. Mix slowly, but thoroughly, using the dough hook.
2. Add the butter pieces a few at a time and mix in slowly on low speed. This should take about 20 minutes, and the dough should be hanging onto the dough hook.

3. Place the dough in three 1 L (3–4 c) baking pans. They should be a little less than half full, as the dough will rise by about two-thirds.
4. Cover with plastic wrap and place in a cool spot for 12 hours. Then let the dough rise at room temperature for 6 hours.
5. Preheat the oven to 160°C (320°F). Bake the bread for 20–25 minutes.
6. Before serving, cut thin slices of the brioche and toast them until they are golden and crisp.

Sesame paste

200 g (⅞ c) black sesame seeds
½ dL (⅕ c) sesame oil
a little salt

7. Toast the sesame seeds light in a dry skillet. Blend with the sesame oil and a little salt in a thermo blender at 70°C (160°F) for 10 minutes. Alternatively, one can use about 250 g (½ lb) of ready-made sesame paste.

White chocolate cream

1 dL (⅖ c) heavy cream
1 dL (⅖ c) full-fat milk
3 egg yolks
20 g (⅔ oz) sugar
50 g (1⅔ oz) good-quality white chocolate
siphon flask and 1 cartridge
50 g (1¾ oz) very blue Roquefort, frozen

8. Heat the cream and milk in a saucepan. Mix together the egg yolks and sugar in a bowl. Stir the eggs slowly into the milk mixture a little at a time, while increasing the heat slowly. Be careful not to let the eggs scramble.
9. Break the chocolate into pieces and place in a bowl.
10. Pour the milk mixture over the chocolate, stir gently until the chocolate has dissolved, and then refrigerate until chilled.
11. Pass the cold mixture through a sieve into the siphon flask, insert the cartridge, and keep cold until ready to serve.

To serve: Pipe the airy white chocolate mixture from the siphon flask into deep bowls, drizzle the sesame paste on top, and grate the blue cheese over it. Finish by crumbling broken pieces of brioche over the top.

▸ White chocolate cream, black sesame seeds, Roquefort, and brioche with nutritional yeast.

*The pleasure of
dining belongs to all
ages, to all conditions, to
all countries, and to all times;
it mingles with all the other
pleasures, and remains at
the end to console us for
the loss of the rest.*

Le plaisir de la table est de tous les âges, de toutes les conditions, de tous les pays et de tous les jours; il peut s'associer à tous les autres plaisirs, et reste le dernier pour nous consoler de leur perte.

Jean Anthelme Brillat-Savarin (1755–1826)

Umami and wellness

Food with umami can often be prepared with significantly less salt, sugar, and fat without sacrificing the delicious taste of the resulting dish. Salt, in particular, is frequently applied too liberally in order to compensate for ingredients that are insipid or unpalatable. In many cases, its use can be reduced by as much as a half by incorporating foodstuffs with umami into the recipe. The fifth taste spurs the appetite, an attribute that can be exploited to advantage in caring for the sick and the elderly, who may have lost interest in eating. At the same time, however, umami promotes satiety, which helps to curb overeating by those who are inclined to overindulge. Either way, adopting a diet that has an abundance of umami may be a way for modern humans to eat in a healthier manner and to adjust their caloric intake to suit the needs of their bodies.

UMAMI AND MSG: FOOD WITHOUT 'CHEMICALS'

On virtually any given day of the week, one can see advertisements in newspapers and magazines for products—usually prepared foods, cosmetics, and textiles—that purport to be free of 'chemicals.' Whether we like it or not, however, at a fundamental level all material substances are made up of chemicals. So the advertising slogans are complete nonsense and shift the focus away from the more central questions of whether the items are wholesome, nutritious, safe to use, and sustainable. To a certain extent, the effectiveness of such advertising depends on ignorance or scientific illiteracy and, in the worst-case scenario, on a willful denial of established facts. The clear implication is that chemistry and chemicals pose a danger to humans; the subtext is that those products that are derived as directly from nature as possible are better for us.

Unfortunately, this is a distortion of reality. For example, a very significant proportion of natural ingredients sourced from plants and fungi are, to varying degrees, poisonous. There is nothing strange about this. Many living organisms either contain certain chemicals in their bodies or are able to secrete them as part of their survival strategy to protect themselves from predators. In contrast, many synthetic products are harmless and safe, precisely because they are made under well-controlled conditions from chemically pure ingredients that are known to be harmless.

For a certain segment of the population, food without chemicals basically means food without additives. But this raises two important questions: What constitutes an additive, and to what extent should it be considered natural? By definition, table salt (NaCl) is regarded as a food, regardless of how it has been produced, and therefore it does not have to be declared as an additive. The same holds true for other household staples, like vinegar. This is a matter of history and tradition. In contrast, monosodium glutamate (MSG), regardless of how it was produced, whether naturally or synthetically, is defined as a food additive by the governments of some countries and must be identified as such. In the United States, MSG is declared GRAS (generally recognized as safe), but foods containing added MSG must list it on the ingredient panel on the packaging. In Australia, New Zealand, and the European Union, MSG has to be listed as a food additive.

As shown in the tables at the back of the book, many so-called natural foodstuffs contain a large amount of MSG, either when raw or as a result of cooking, fermenting, drying, or aging. It is therefore misleading to talk about foods to which MSG has been added as foods with chemicals.

It is also important to keep in mind that too much MSG has the effect of making food less palatable. In a sense, then, it is considered a self-limiting substance, and there is a general tendency on the part of food processors to use the least amount necessary to maximize taste.

Generally speaking, regulatory authorities have not issued rulings on the subject of safe daily limits for the intake of MSG, as no one has been able to determine exactly what such a limit might be. Some researchers have cautiously suggested that it could be set at 2.1 grams per kilogram of body weight. This quantity is so relatively large that it is highly unlikely that anyone would ingest even a tiny fraction of this amount.

Regrettably, even though MSG is the most intensely researched substance found in foods, it is virtually impossible, even in a well-informed society, to dislodge what amounts to a superstition; namely, that food with MSG is food that is laced with dangerous chemicals.

UMAMI SATISFIES THE APPETITE

Umami can help to regulate appetite. When one has ingested enough umami substances, the desire to eat more attenuates and food intake is limited to a level that matches the body's nutritional needs. At the same time, signals are sent from the receptors on the tongue to the stomach and the pancreas to forewarn them that protein-rich food is on its way. In this way, the presence of umami promotes better digestion of the proteins. The effect of the ingestion of umami on appetite and digestion illustrates a general phenomenon called homeostasis. Homeostasis describes a balance in which opposing tendencies interact to produce a form of self-regulation that maintains an equilibrium. Food with umami can, therefore, serve as a good weapon in the global fight against obesity.

**WHY DOES UMAMI MAKE US FEEL FULL?
THE 'BRAIN' IN THE STOMACH**

Recent research may be well on the way to finding a physiological explanation for why umami can act as an appetite regulator. The stomach and intestines have a well-developed nervous system, which, to a certain extent, functions autonomously and independently from taste perception, but which is also linked to the brain via a certain cranial nerve, the vagus nerve. One can think of this network of nerves in the stomach as a sort of mini-brain that is responsible for peristalsis in the intestines, digestion, and part of the immune system. This 'stomach brain' in humans has just as many nerve cells as a cat brain and just as many different types of nerve cells as the human brain. It is this 'stomach brain' that discerns when we have had enough to eat. But sometimes will power causes the actual brain to take control and ignore the signal, and we continue to eat.

A glutamate receptor that has been found in the stomach of research animals resembles the one found in the taste buds on the tongue. Experiments showed that an increase in glutamate in the stomach, by binding to this receptor, stimulates the vagus nerve, which in turn sends a signal to the brain that the fundic glands in the stomach sac need to secrete more of certain enzymes (proteases), which break down proteins. In this way the glutamate may act as a trigger that sets in motion the chain of events alerting the stomach to prepare for the imminent arrival of protein-rich foods that need to be digested. Perhaps in the course of evolution this effect has been partly responsible for the extent to which we like the taste of glutamate. Apparently, other amino acids are not able to activate this signal.

UMAMI FOR A SICK AND AGING POPULATION

A long stay in the hospital or certain debilitating medical treatments can leave seriously ill people in a very weakened state. Even the act of eating can pose problems for them if they lack physical strength to do so or if they have lost their appetite. By turning our attention to how to optimize umami in the food, we could design small, nutritious portions that have more appeal. One way would be to reduce drastically the amount of cooking liquid to result in a more concentrated taste.

In many parts of the world, the population is aging. Even though these older people may remain active and healthy into advanced age, they will inevitably suffer from illnesses and aches and pains that reduce the quality of life. In addition, the senses become less acute, particularly over the age of seventy. It is not just eyesight that is affected but also the ability to taste and smell. This can lead to loss of appetite because the food seems to be less appealing and is judged to be less tasty, with the result that older people may eat too little or seek out the wrong foods. The consequences are malnutrition or insufficient caloric intake, a weakened immune system, and greater risk of illness. An associated problem is that bland food does not stimulate the secretion of sufficient saliva, which leads to poorer digestion and an attendant diminished uptake of nourishing substances. These difficulties affect not only the old who are living on their own but also those who are in nursing homes or hospitalized.

Experiments have demonstrated that many of these problems can be alleviated by augmenting the umami content of the food, either by adding MSG directly or simply by preparing it with ingredients that are able, from the outset, to impart more savoriness. Tests have also shown that umami has a much greater ability than sourness to stimulate secretion of saliva during an ensuing period of two to ten minutes. More saliva increases the capacity to dissolve more taste substances, which intensifies their taste, and it becomes easier to chew and swallow the food. As a consequence, the elderly are more willing to eat, leading to improvement in their nutritional state and boosting their resistance to disease.

Another important consideration is that adding umami allows for a reduction in the salt content, which is good news for those among the elderly who suffer from high blood pressure. There are already moves to limit the amount of salt in institutional food, but this leads to many complaints about how it tastes. By placing greater emphasis on how to add more umami to the food, it is possible to get by with less salt without sacrificing palatability.

There are very few places where umami has been introduced systematically as a means of improving institutional food. Even in Japan, where umami was first given a name, investigations have shown that the salt content of the food for elderly patients in nursing homes is regulated within a relatively narrow range (361–1,516 mg/100 g), whereas the glutamate content varies somewhat arbitrarily (16–697 mg/100 g). Seemingly, even in Japan, umami has not yet been adopted as a central parameter for the preparation of food for older patients.

It would certainly appear that there is a vast, unexploited potential for making informed use of the fifth taste in many of the kitchens where food is prepared for the sick and the old.

UMAMI FOR LIFE
From the day we are born until the day we die, our lives as *Homo sapiens* are inseparably bound up with our relationship to food and its taste. Our health and survival are dependent on having a balance of different and sufficient nutrients. The quality of our lives is tied to the palatability of the food and the harmonious taste impressions we experience.

From prehistoric times, our species has had an instinct to seek out food that has a high energy content and is nutritious. In a healthy person, this instinct is underpinned by self-regulating mechanisms that help to control our weight and determine our overall physical well-being. In this context, an understanding of umami becomes a central concept for enhancing wellness and enjoyment of life, from start to finish.

It behooves us to become better at acknowledging the meaning of umami at every stage of our lives. There is a reason why it is found in mother's milk; there is a reason why, throughout our lives, we collect fond memories of delicious meals; and there is a good reason why we need to make sure that the sick and elderly among us have access to food that tastes better and has more umami.

Epilogue: Umami has come to stay

As we have seen throughout this book, umami is a relatively new label for a taste that, for possibly the past 1.9 million years, has been an integral aspect of the food of modern humankind and its ancestors. It is an attribute of nutritious food and in this way has steered our preference for food with that particular taste. The taste is intensified when we work with the raw ingredients in certain ways, which have been refined in the course of millennia and which are the very heart of our food cultures, culinary skills, and gastronomy. Virtually all the cuisines in the world seem to strive to impart umami, each with its typical and regional raw ingredients and centuries-old techniques. Of all the techniques, cooking, aging, and fermenting are best able to draw out umami.

Generations of housewives, cooks, and chefs have known intuitively how to elicit umami and that it is indispensable. In more recent times, food manufacturers, gourmets, and innovative chefs have become aware of its synergistic effect and have started to tap into its potential in a rational, creative way. Nevertheless, many of us have not yet gained an easy familiarity with the word umami as an expression to describe savoriness in our raw ingredients, our food, our meals, and our food cultures.

Science has taught us which substances in the raw ingredients can help to impart umami, and, armed with this knowledge, we are better able to understand why food has umami tastes and, just as important, what we have to do to enhance them. We now also know that what characterizes umami is the multiplier effect. This taste comes fully into its own only with the help of an intimate interaction, a synergy between two types of substances, glutamate and ribonucleotides. An awareness of which raw ingredients are sources of these two substances allows us to sharpen our insight into how we can prepare more delicious meals. While this will naturally be of great value in the field of advanced gastronomy, it is of equal importance in our own kitchens, where we can use it to real advantage, even with simple techniques and local ingredients.

It is our contention that delicious effects attributable to umami can be combined with good eating habits and overall wellness. Deliciousness

◀ Sun-ripened tomatoes are rich in glutamate and adenylate and hence contribute perfect umami synergy.

can be a source of greater enjoyment and satisfaction in a meal without leading to gluttony. Current estimates indicate that more than one out of every five people in the world is overweight; obesity is rapidly becoming a global epidemic. In its wake comes a long list of illnesses, not least of which are cardiovascular disease and diabetes. To a large extent, this increase in obesity is due to a change in dietary habits and a lack of physical exercise. The change in dietary habits is often caused by lack of information about raw ingredients, little experience in food preparation, and an ignorance of the importance of a real food culture and of sitting down together at mealtimes. We have seen the quality of the raw ingredients and the time needed to prepare them carefully being forced, time and again, to give way to quantity and convenience. Knowledge of umami can help to counteract these trends by inspiring us to produce healthier, tastier meals with reduced salt, sugar, and fat contents and to use foodstuffs more fully, with less waste.

In many Western countries, the diet is out of balance when it comes to fruits and vegetables. The recommended daily intake is 600 grams, but many people have problems coming near that quota, especially when it comes to vegetables. Some vegetables have very few umami substances and are not very palatable, especially if they are eaten raw. Once again, we could turn to the Japanese 'enlightened kitchen' to devise ways of serving delicious food made with vegetables. The secret is to use dashi to bring umami to the vegetables. As we have seen in this book, it is possible to imbue otherwise bland ingredients, such as green salads and certain vegetables, with a savory taste by combining them with ingredients that are a source of either basal or synergistic umami.

We consider umami to be the central point around which the circle of deliciousness revolves and are convinced that it deserves a place of honor in all the food cultures of the world. Let us make the appreciation of umami our challenge and the discovery of savoriness our mission. We and our children should feel free to experiment in our own kitchens without being bound by the limits imposed by traditions or predetermined by the mass production of foods. It is our hope that every generation will build on the past to interpret umami for itself and seek it out in fresh, new ways.

Armed with its official recognition by the scientific community as a true basic taste and bolstered by our own heightened awareness of the role it plays in our food, umami has joined the ranks of the indispensible culinary tools—it has come to stay.

▸ Oysters are rich in glutamate and hence umami taste.

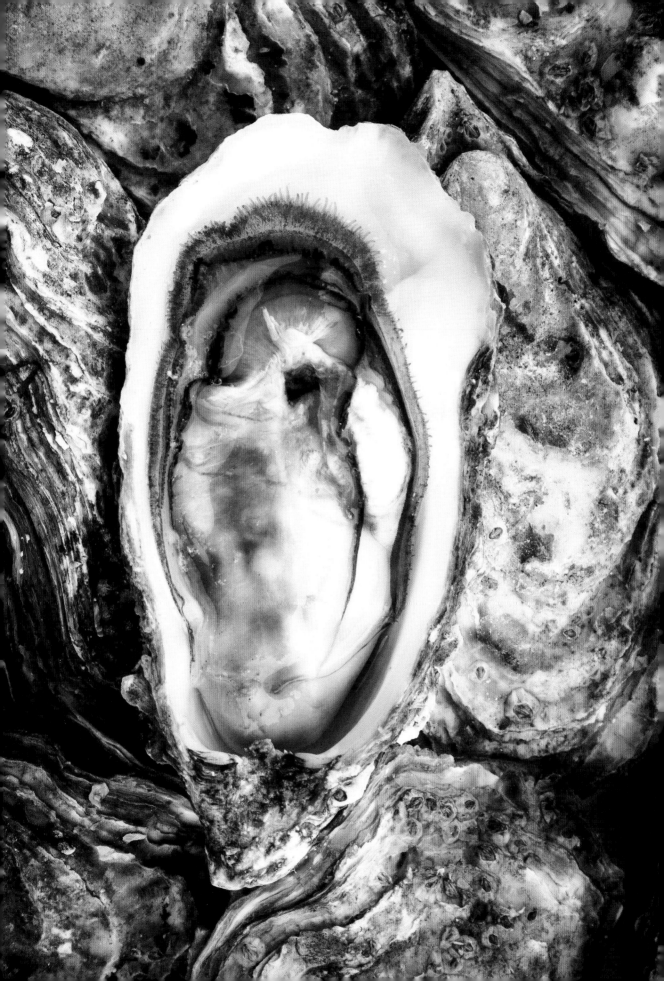

*Of course
people have made
and enjoyed well-seasoned
dishes for thousands of years
with no knowledge of molecules.
But a dash of flavor chemistry
can help us make fuller use of our
senses of taste and smell, and
experience more—and find
more pleasure—in what
we cook and eat.*

Harold McGee (1951–)

Technical and scientific details

UMAMI AND THE FIRST GLUTAMATE RECEPTOR
With the discovery in 2000 of the first umami receptor, *taste*-mGluR4, scientists were finally in a position to investigate the molecular basis of umami, which Professor Ikeda, as early as 1908, had already determined could be elicited by glutamate.

The peculiar name of the receptor, *taste*-mGluR4, tells us that it is related to the already known mGluR4 in the brain, which is sensitive to glutamate when it acts as a neurotransmitter. The difference between *taste*-mGluR4 and mGluR4 is that the former is a truncated version of the latter; the part of the taste receptor that projects from the taste cell is only about half as big as that of *brain*-mGluR4. As the outermost part of the receptor molecule is exactly the site where glutamate is bound, the consequence is that the taste receptor is much less sensitive than its larger counterpart in the brain. Actually, *taste*-mGluR4 is more than one hundred times less sensitive than *brain*-mGluR4. It is, however, not important for the taste receptors to be nearly as sensitive as those in the brain because the concentration of glutamate in food is much greater than the concentration of glutamate in the neural cells.

Not all the details of how the taste receptor *taste*-mGluR4 works are known, but seemingly it is activated when glutamate is bound to the part of the protein that projects from the membrane. This part is very flexible, almost like a hinge. As soon as the glutamate molecule binds, the hinge partly closes around it, thereby holding it more strongly. It turns out that this closing mechanism is also the secret behind the synergistic interaction of the ribonucleotides inosinate and guanylate, which enhances umami. We will return to this later.

When glutamate is bound to the binding site in the receptor, a signal is sent through the protein, so that a G-protein is bound to the receptor on the other side of the membrane, which faces toward the inside of the cell. This indicates that the signal has been received and triggers a cascade of sequential biochemical processes: Certain ion channels open, which results in a drop in the membrane's electrical potential, which again sends an impulse through the nerve cell to the brain. The taste impression has arrived at its destination.

Schematic illustration of a receptor embedded in the cell membrane of a sensory cell.

YET ANOTHER RECEPTOR FOR UMAMI: THIS IS HOW IT WORKS
Two years after the identification of the *taste*-mGluR4 receptor, scientists found a second, more complicated type of umami receptor. It is from a family of receptor cells, called T1R-receptors, found uniquely in the sensory cells. And like the *taste*-mGluR4 receptors, they are distantly related to the glutamate receptors in the brain.

This new receptor reveals a little more about the nature of umami and how the perception of this taste is possibly related to that of sweetness. Furthermore, as discussed below, this receptor is also, and quite surprisingly, a key to understanding synergy in our experience of umami; that is, the way in which some ribonucleotides enhance the perception of glutamate.

In 2002, two groups of researchers working independently discovered that members of the T1R family of receptors could assemble into pairs, forming two related complexes, T1R1/T1R3 and T1R2/T1R3, which are sensitive to umami and sweet tastes, respectively. The similarity between them may be an indicator that the biochemical pathways for the recognition of sweetness and umami are very closely connected. In fact, they may be so intimately related that it would explain why rats experience umami as a sweet taste and stop eating sugar when they have ingested too much glutamate. It may also explain why umami can enhance the perceived sweetness of some foodstuffs.

In rats and mice, the T1R1/T1R3 pair is sensitive to a broad spectrum of left-turning amino acids, but in humans it responds most strongly to glutamate and aspartate. On the other hand, this receptor complex is not sensitive to right-turning amino acids or other substances. Only about 70 percent of the T1R1 in rodents is identical to that in humans. The binding site for glutamate in the T1R1/T1R3 receptor is found in the T1R1 part, and it has the same molecular structure as in the mGluR4 and *taste*-mGluR4 receptors. In contrast to *taste*-mGluR4, however, T1R1/T1R3 is not sensitive to L-aspartate, which has only a little umami.

It is also interesting that, although glutamate binds exclusively to T1R1, the receptor system works only if T1R1 is paired with T1R3 and if T1R3 is intact. Genetic variations expressed in the T1R3 receptor can, for example, have an effect on the ability to detect umami. In addition, substances that bind to T1R3, and in this way suppress sweetness, can also suppress umami. An example of this effect is the commercial attempt to use the substance lactisole as a taste modifier to diminish the taste of certain natural and artificial sweeteners. Because a lactisole molecule binds itself to T1R3, it has an effect on the sensitivity of the receptor pair T1R2/T1R3 to sweet things. But, in so doing, lactisole also has an impact on the function of T1R1 and, as a totally unforeseen side effect, also suppresses umami.

An essential feature of the discovery of the T1R1/T1R3 receptor complex as it relates to umami is that its sensitivity to glutamate is strengthened in a very robust way by inosinate and guanylate, a very important aspect of umami synergy. In contrast, the receptor cannot be made to respond to right-turning amino acids with the help of inosinate. As well, the receptor is not sensitive to inosinate on its own, which is also characteristic of umami.

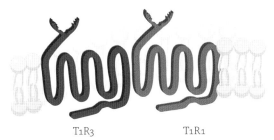

T1R3   T1R1

Schematic illustration of the umami receptor T1R1/T1R3 embedded in the cell membrane of a sensory cell.

In summary, T1R3 is a common partner for the perception of both sweetness and umami. This suggests that in forming pairs with T1R1 and T1R2, respectively, T1R3 itself is neutral with respect to tastes but nevertheless has an effect on the sensitivity of the coupled pairs to these two tastes.

More recently, yet another mGlu glutamate receptor, which is related to *taste*-mGluR4, has been discovered. This brings the total number of receptors for umami to at least three, the two mGlu receptors and T1R1/T1R3. Although it is not known for sure, there is some indication that the signaling pathways for the three types of receptors are different, even though they may involve the same G-proteins (possibly gustducin). It has been suggested that mGluRs and the T1R1/T1R3 pair have different functions in the perception of umami. According to this theory, T1R1/T1R3 plays a major role on the front part of the tongue in the preferential selection of food that has umami, whereas the mGlu receptors are important on the back part of the tongue with regard to discriminating between umami and other tastes.

UMAMI SYNERGY: IT FUNCTIONS LIKE A VENUS FLYTRAP
The discovery of how the taste receptors T1R1 and T1R3 work in tandem in the combination T1R1/T1R3, which can be activated by the substances that synergistically induce umami, paved the way for further research. The scientists were at last on the way to finding out what cooks the world over have known for a very long time, namely, that meat soup with vegetables tastes good. The meat contributes inosinate and the vegetables have glutamate. Or what the Japanese have also known for centuries: Dashi is best when konbu, containing glutamate, is combined with *katsuobushi* or shiitake, which contribute inosinate or guanylate, respectively.

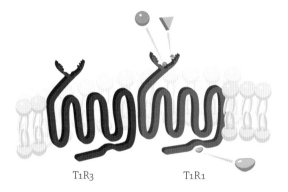

Schematic illustration of the umami receptor T1R1/T1R3 embedded in the cell membrane of a sensory cell. The T1R1 part binds a glutamate ion (the green sphere) and a nucleotide (the blue triangle). This triggers a signal through the receptor that a G-protein (the blue 'jelly bean' shape) is to be bound on the other side of the membrane.

T1R3   T1R1

Inosinate and guanylate do not activate T1R1/T1R3 by themselves, but can do so in conjunction with glutamate. Conversely, other ribonucleotides that do not bring out umami have no effect on the receptor. From recent research it is possible to explain this synergy with the help of a mechanism that resembles the trapping mechanism in the Venus flytrap, an unusual carnivorous plant.

The activity all takes place in the T1R1 part of the receptor complex, which has a hinge-like structure. L-glutamate binds near the place where the hinge will bend. As in the Venus flytrap, this causes the hinge to snap shut, trapping the glutamate molecule securely. Inosinate or guanylate, on the other hand, binds to a place at the edge of the hinge, where there is a spot that resembles a cleft. This increases the trapping power of the hinge and results in an even stronger binding of the glutamate molecule. Stabilizing the glutamate in this way is equivalent to making the receptor more sensitive with respect to glutamate. On the other hand, neither of the two ribonucleotides would be able to stimulate the receptor on its own in the absence of glutamate. Chemists refer to this effect as allostery, which in this case means that the effect of the protein's function is dependent on something that happens in a place on the protein other than its active binding site.

Recent research has shown that the allosteric action is manifested in the dynamics of the Venus flytrap: Without bound glutamate, the flytrap is very dynamic; when glutamate binds, the dynamics are slowed down significantly; and when a nucleotide also binds, the dynamics are extremely slow.

Furthermore, it turns out that glutamate is also involved in stabilizing the actual pair formation of T1R1/T1R3. The whole process, which results in umami taste, is therefore a truly cooperative effort that depends on a number of separate entities coming together to work as a team.

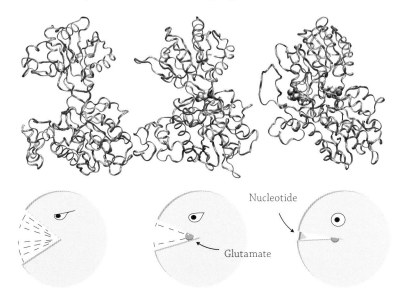

The molecular mechanism of the umami receptor (T1R1/T1R3), illustrating the synergistic action by simultaneous binding of glutamate and a 5'-ribonucleotide (in this case guanylate).

Schematic comparison between the dynamic functioning of the umami receptor and Pac-Man. Without glutamate in his throat, Pac-Man's jaws are very dynamic; when glutamate is bound, the jaw motion is slowed down; when, in addition, a nucleotide binds at his lips, the dynamic motion stops altogether and he is silent.

THE TASTE OF AMINO ACIDS

The twenty different naturally occurring free amino acids can be divided into groups according to their tastes, as shown in Table 1. There are two amino acids that are considered to be tasteless or neutral in taste, aspartic acid and asparagine. The salt of aspartic acid, aspartate, has a slight umami taste. Some sweet amino acids, such as lysine and proline, taste bitter in large quantities. One of the amino acids, methionine, tastes both bitter and sulfurous. The taste of the different amino acids can vary with pH.

Molecular structures of MSG (monosodium glutamate) and MSA (monosodium aspartate).

MSG   MSA

TABLE 1: THE TASTE OF AMINO ACIDS

Umami	Bitter	Sweet	Sour	Neutral
aspartate	arginine	alanine	cysteine	asparagine
glutamate	histidine	glutamine		aspartic acid
(glutamic acid)	isoleucine	glycine		
	leucine	lysine		
	methionine	proline		
	phenylalanine	serine		
	tryptophan	threonine		
	tyrosine			
	valine			

Source: Belitz et al. *Food Chemistry*. Springer, New York, 2004, p. 34, Table 1.12.

TASTE THRESHOLDS FOR UMAMI

Table 2 shows the taste thresholds as a percentage by weight for substances that impart an umami taste in pure water. The given values are subject to a considerable degree of uncertainty.

TABLE 2: TASTE THRESHOLDS FOR UMAMI

MSG	0.03%
IMP	0.012%
GMP	0.0035%
MSG+IMP	0.0001%
MSG+GMP	0.00003%

MSG = monosodium glutamate; IMP = inosine-5'-monophosphate; GMP = guanosine-5'-monophosphate.
Source: Maga, J. A. Flavor potentiators. *Crit. Rev. Food Sci. Nutr.* **18**, 231–312, 1983.

CONTENT OF GLUTAMATE AND 5'-RIBONUCLEOTIDES IN DIFFERENT FOODS

Tables 3 to 11 list the content of bound and free glutamate, as well as of 5'-ribonucleotides, in a variety of raw ingredients and prepared products. The data is drawn from a number of sources, and it should be stressed that there can be considerable variation in currently available values for the same foodstuff. In addition, the content of these substances can, to a large extent, depend on the specifics of a given sample, such as the species, place of origin, degree of ripeness, storage conditions, and so forth. Regrettably, there is often no information in the technical and scientific literature regarding certain raw ingredients or processed foods.

There can also be considerable variation between the content of free and of bound glutamate in a particular raw ingredient. The amount of free glutamate is indicative of the foodstuff's innate ability to impart umami taste. On the other hand, the bound glutamate content is indicative of the ingredient's potential to develop more umami taste substances if it is treated in such a way that the proteins are broken down into free amino acids. Examples of processes used to do so include cooking, fermenting, curing, dehydrating, and marinating.

TABLE 3: BOUND AND FREE GLUTAMATE IN A SELECTION OF RAW INGREDIENTS

Food category	Bound glutamate (mg/100 g)	Free glutamate (mg/100 g)
Meat		
Beef	2,846	33
Pork	2,325	23
Poultry		
Duck	3,636	69
Chicken	3,309	44
Eggs	1,583	23
Fish		
Cod	2,101	9
Mackerel	2,382	36
Salmon	2,216	20
Vegetables		
Green peas	5,583	106
Corn	1,765	130
Carrot	218	20
Spinach	289	48
Tomato	238	140
Potato	280	102
Milk and cheese		
Parmigiano-Reggiano	9,847	1,680
Cow's milk	819	1
Human breast milk	229	19

Source: Ninomiya, K. Natural occurrence. *Food Rev. Int.* **14**, 177–211, 1998.

TABLE 4: COMPARISON OF AMINO-ACID CONTENTS IN DASHI AND VARIOUS SOUP STOCKS

The table lists data for the content of umami free amino acids (glutamic acid and aspartic acid) and a sweet amino acid (alanine) in different preparations of dashi and soup stock. The recipe calls for 1.8 L water and 30 g seaweed. The seaweed is soaked in the water and extracted at 60°C for 1 hour. The *Rausu-konbu* dashi and the dulse dashi are prepared by using about twice as much seaweed per liter of water as for the *Rishiri-konbu* dashi and *ichiban* dashi.

Dashi or soup stock	Glutamic acid mg/100 g	Aspartic acid mg/100 g	Alanine mg/100 g
Rishiri-konbu dashi	22	16	1
Rausu-konbu dashi (traditional)	100	60	7
Rausu-konbu dashi (*sous-vide*)	145	85	20
Ichiban dashi	25	18	4
Dulse dashi	40	27	25
Western chicken stock	18	6	11
Chinese *tang* (chicken base)	14	4	8

Sources: Mouritsen, O. G., L. Williams, R. Bjerregaard & L. Duelund. Seaweeds for umami flavour in the New Nordic Cuisine. *Flavour* **1**:4, 2012; Ninomiya, K. Unpublished data from the Umami Information Center; Kurihara, K. Glutamate: from discovery as a food flavor to role as a basic taste. *Am. J. Clin. Nutr.* **90**, 719S-722S, 2009; Ozawa, S., H. Miyano, M. Kawai, A. Sawa, K. Ninomiya, K. Mawatari & M. Kuroda. Changes of free amino acids during cooking process of chicken consommé (in Japanese) presented at the 58th Congress of the Japanese Society for Nutrition and Food Science, Sendai, May 21–23, 2004, p. 322; Ozawa, S., H. Miyano, M. Kawai, A. Sawa, K. Ninomiya, K. Mawatari & M. Kuroda. Changes of free amino acids during cooking process of Chinese chicken bouillon (in Japanese) presented at the 59th Congress of the Japanese Society for Nutrition and Food Science, Tokyo, May 13–15, 2005, p. 322.

TABLE 5: FREE GLUTAMATE IN RAW INGREDIENTS

Food category	Free glutamate (mg/100 g)	Food category	Free glutamate (mg/100 g)
Meat and poultry		**Vegetables**	
Ham (air-dried)	337	Tomato (sun-dried)	648
Duck	69	Tomato	200
Chicken	44	Potato (cooked)	180
Beef	33	Potato	102
Pork	23	Corn	130
Eggs	23	Broccoli	115
Lamb	8	Green peas	106
		Lotus root	103
Fish and shellfish		Garlic	99
Anchovies (marinated)	1,200	Chinese cabbage	94
		Soybeans	66
Sardines	280	Onion	51
Squid	146	Cabbage	50
Scallops	140	Green asparagus	49
Sea urchin	140	Spinach	48
Oysters	130	Lettuce	46
Mussels	105	Cauliflower	46
Caviar	80	White asparagus	36
Alaska crab	72	Carrot	20
Sardines (dried, *niboshi*)	50	Marrow	11
		Green bell pepper	8
Shrimp	40	Cucumber	1
Mackerel	36		
Dried bonito	36	**Milk**	
Dried tuna	31	Human breast milk	19
Salmon roe	22	Goat's milk	4
Salmon	20	Cow's milk	1
Crab	19		
Cod	9	**Fungi**	
Lobster	9	Shiitake (dried)	1,060
Herring	9	Shiitake	71
		Button mushroom	42
Tea		Truffle	9
Green tea	450		
Green tea (roasted)	22		

Food category	Free glutamate (mg/100 g)	Food category	Free glutamate (mg/100 g)
Fruits and nuts		**Cheese**	
Walnuts	658	Parmigiano-Reggiano	1,000–2,700
Strawberries	45	Roquefort	1,280
Apple juice	21	Gruyère de Comté	1,050
Pear	20	Stilton	820
Avocado	18	Cabrales (goat's milk blue cheese)	760
Kiwifruit	5		
Red wine grapes	5		
Grapefruit	5	Danish blue	670
Apple	4	Gouda	460
		Camembert	390
Dried seaweeds		Emmenthal	308
Konbu (*Saccharina japonica*)	1,400–3,200	Cheddar	182
Nori (*Porphyra yezoensis*)	1,378		
Wakame (*Undaria pinnatifida*)	9		

Sources: Ninomiya, K. Umami: a universal taste. *Food Rev. Int.* **18**, 23–38, 2002; Ninomiya, K. Natural occurrence. *Food Rev. Int.* **14**, 177–211, 1998; http://www.umami-info.com; http://www.msgfacts.com; Özden, Ö. Changes in amino acid and fatty acid composition during shelf-life of marinated fish. *J. Sci. Food Agric.* **85**, 2015–2020, 2005; Löliger, J. Function and importance of glutamate for savory foods. *Amer. Soc. Nutr. Sci.* **130**, 915S-920S, 2000; Giacometti, T. Free and bound glutamate in natural products. *Glutamic Acid: Advances in Biochemistry and Physiology* (L. J. Filer, Jr. et al., eds.) Raven Press, New York, s. 25–34, 1979; Komata, Y. Umami taste of seafoods. *Food Rev. Int.* **6**, 457–487, 1990; Maga, J. A. Flavor potentiators. *Crit. Rev. Food Sci. Nutr.* **18**, 231–312, 1983.

TABLE 6: FREE GLUTAMATE IN FERMENTED FOOD PRODUCTS

Food category	Free glutamate (mg/100 g)
Soy sauce	
Korea	1,264
China	926
Japan	782
Fish sauce	
Japan	1,383
Vietnam	1,370
China	828
Modern *garum*	
Garum	623
Quick-and-easy *garum*	217
Fermented beans	
West African *soumbala* (*Parkia biglobosa*)	1,700
Chinese *douchi* (soybeans)	1,080
Miso	500–1,000
Tempeh	985
Japanese *nattō* (soybeans)	136
Fermented rice	
Sake	186

Sources: Ninomiya, K. Umami: a universal taste. *Food Rev. Int.* **18**, 23–38, 2002; Ebine, H. Miso preparation and use. *Food Uses of Whole Oil and Protein Seeds*. (E. W. Lusas, D. R. Erickson & W.-K. Nip, eds.) American Oil Chemists' Society, Champaign IL, pp. 131–147, 1986; Murata, K., H. Ikehata & T. Miyamoto. Studies on the nutritional value of tempeh. *J. Food. Sci.* **32**, 580–586, 1967; www.kikumasamune.co.jp; glutamate content of *garum* and quick-and-easy *garum* measured by Niels O. G. Jørgensen.

TABLE 7: INCREASE OF FREE GLUTAMATE IN A RIPENING TOMATO

The table lists the free glutamate content (mg/100 g) in a field tomato, from the green to the overripe stage. It should be noted that the content varies from one part of the tomato to another, with the innermost part often having up to five times as much as the outer surface. Free glutamate content also varies by species, with cherry tomatoes having the most.

Green	Ripe green	Partly red	Pale red	Red	Fully ripe	Over-ripe
20	21	30	74	143	175	263

Source: Ninomiya, K. Natural occurrence. *Food Rev. Int.* **14**, 177–211, 1998.

TABLE 8: INCREASE OF FREE GLUTAMATE IN AGING CHEDDAR

The table lists the free glutamate content (mg/100 g) in cheddar cheese from the time it is made until it has been aged for eight months.

Aging time (months)	0	1	2	3	4	5	6	7	8
Free glutamate (mg/100 g)	11	22	36	54	78	112	121	160	182

Source: Ninomiya, K. Natural occurrence. *Food Rev. Int.* **14**, 177–211, 1998.

TABLE 9: INCREASE OF FREE GLUTAMATE IN A CURING HAM

The table lists the progression, over a period of eighteen months, of the free glutamate content (mg/100 g) in ham that is being air dried. The curing process involves a complicated cycle of salting and drying at different temperatures and humidity levels. The ham loses about 30 percent of its initial water content.

Curing time (months)	0	1	2	4	6	12	18
Free glutamate (mg/100 g)	6	11	35	46	142	207	337

Source: Ninomiya, K. Natural occurrence. *Food Rev. Int.* **14**, 177–211, 1998.

TABLE 10: 5'-RIBONUCLEOTIDES IN RAW INGREDIENTS

Content of free 5'-ribonucleotides in a variety of raw ingredients. The data is drawn from different sources, and there can be considerable variation in available values for the same foodstuff. Moreover, the content can, to a large extent, depend on the specifics of a given sample, such as the species, place of origin, degree of ripeness, and storage conditions.

Food category	IMP (mg/100 g)	GMP (mg/100 g)	AMP (mg/100 g)
Fish and shellfish			
Katsuobushi (dried bonito)	687		52
Niboshi (dried sardines)	863		
Anchovy paste	300	5	
Sardines	193		6
Scallops	-	-	172
Sea urchin	2	2	10
Mackerel	215	-	6
Tuna	286	-	6
Salmon	154	-	6
Lobster	-	-	82
Cod	44		23
Shrimp	92	-	87
Squid	-	-	184
Crab	5	5	32
Vegetables and fungi			
Tomato (sun-dried)	-	10	
Tomato	-	-	21
Potato (cooked)	-	2	4
Green peas	-	-	2
Shiitake	-	16–45	
Shiitake (dried)	-	150	
Matsutake	-	65	
Morel (dried)	-	40	
Enokitake		22	
Porcini mushrooms (dried)	-	10	
Oyster mushrooms (dried)	-	10	
Green asparagus	-	-	4

Food category	IMP (mg/100 g)	GMP (mg/100 g)	AMP (mg/100 g)
Meat and fowl			
Chicken	201	5	13
Pork	200	2	9
Beef	70	4	8
Seaweeds			
Nori (*Porphyra yezoensis*)	9	5	52
Milk			
Human breast milk	0.3		

IMP = inosine-5'-monophosphate; GMP = guanosine-5'-monophosphate; AMP = adenosine-5'-monophosphate; - = below measurable threshold; blank = no available data.

Sources: Ninomiya, K. Umami: a universal taste. *Food Rev. Int.* **18**, 23–38, 2002; Ninomiya, K. Natural occurrence. *Food Rev. Int.* **14**, 177–211, 1998; Goral, D. M. Flavour-enhancing food additive. US patent 7,510,738B2, 2009 ; Maga, J. A. Flavor potentiators. *Crit. Rev. Food Sci. Nutr.* **18**, 231–312, 1983.

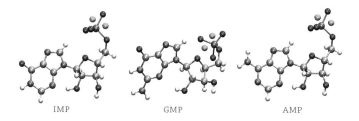

Molecular structures of the nucleotides IMP, GMP, and AMP.

TABLE 11: CHANGES IN CONTENT OF UMAMI SUBSTANCES IN MEAT AND FOWL AFTER SLAUGHTERING

The table lists the content (mg/100 g) of free glutamate and free IMP (inosine-5'-monophosphate) in the period following slaughter, at a storage temperature of 4°C.

	Beef		Pork		Chicken	
	4 days	12 days	1 day	6 days	0 days	2 days
Glutamate	6	10	4	9	13	22
IMP	90	80	260	226	284	231

Sources: Ninomiya, K. Umami: A universal taste. *Food Rev. Int.* **18**, 23–38, 2002; Ninomiya, K. Natural occurrence. *Food Rev. Int.* **14**, 177–211, 1998.

Bibliography

Araujo, I. E. T., M. L. Kringelbach, E. T. Rolls & P. Hobden. Representation of umami taste in the human brain. *J. Neurophysiol.* **90**, 313–319, 2003.

Ardö, Y., B. V. Thage & J. S. Madsen. Dynamics of free amino acid composition in cheese ripening. *Austr. J. Dairy Technol.* **57**, 109–115, 2002.

Bachmanov, A. Umami: Fifth taste? Flavor enhancer? *Perfum. Flavor.* **35**, 52–57, 2010.

Barham P., L. H. Skibsted, W. L. Bredie, M. B. Frøst, P. Møller, J. Risbo, P. Snitkjaer & L. M. Mortensen. Molecular gastronomy: a new emerging scientific discipline. *Chem. Rev.* **110**, 2313–2365, 2010.

Bartoshuk, L. M. The biological basis of food perception and acceptance. *Food Quality and Preference* **4**, 21–32, 1993.

Belitz, H.-D., W. Grosch & P. Schieberle. *Food Chemistry*. 3. Ed., Springer, New York, 2004.

Bellisle, F. Glutamate and the umami taste: sensory, metabolic, nutritional and behavioural considerations. *Neurosci. Biobehav. Rev.* **23**, 423–438, 1999.

Blumenthal, H. *The Fat Duck Cookbook*. Bloomsbury Publ., London, 2008.

Blumenthal, H., P. Barbot, N. Matsushisa & K. Mikuni. *Dashi and Umami: The Heart of Japanese Cuisine*. Eat-Japan, Cross Media Ltd., London, 2009.

Booth, M. *Sushi and Beyond: What the Japanese Know About Cooking*. Jonathan Cape, London, 2009.

Brillat-Savarin, J. A. *The Physiology of Taste*. Penguin, London, 1970.

Cambero, M. I., C. I. Pereira-Lima, J. A. Ordoñez & G. D. Garcia de Fernando. Beef broth flavour: relation of components with the flavour developed at different cooking temperatures. *J. Sci. Food Agric.* **80**, 1519–1528, 2000.

Chandrashekar, J., M. A. Hoon, N. J. Ryba & C. A. Zucker. The receptors and cells for mammalian taste. *Nature* **444**, 288–294, 2006.

Chandrashekar, J., C. Kuhn, Y. Okal, D. A. Yarmolinsky, E. Hummler, N. J. P. Ryba & C. S. Zuker. The cells and peripheral representation of sodium taste in mice. *Nature* **464**, 297–301, 2010.

Chandrashekar, J., D. Yarmolinsky, L. von Buchholtz, Y. Oka, W. Sly, N. J. P. Ryba & C. S. Zuker. The taste of carbonation. *Science* **326**, 443–445, 2009.

Chaudhari, N., A. M. Landin & S. D. Roper. A novel metabotropic glutamate receptor functions as a taste receptor. *Nature Neurosci.* **3**, 113–119, 2000.

Chen, X., M. Gabitto, Y. Peng, N. J. Ryba & C. S. Zuker. A gustotopic map of taste qualities in the mammalian brain. *Science* **333**, 1262–1266, 2011.

Cross Media (ed.) *Umami the World: The Fifth Taste of Human Beings*. 2ed. Cross Media Ltd., Tokyo, 2007.

Curtis, R. *Garum and Salsamenta: Production and Commerce in* Materia Medica. E. J. Brill, Leiden, 1991.

Delwiche, J. Are there 'basic' tastes? *Trends. Food. Sci. Technol.* **7**, 411–415, 1996.

Dermiki, M., R. Mounayar, C. Suwankanit, J. Scott, O. B Kennedy, D. S. Mottram, M. A. Gosney, H. Blumenthal & L. Methven. Maximising umami taste in meat using natural ingredients: effects on chemistry, sensory perception and hedonic liking in young and old consumers. *J. Sci. Food Agric.* **13**, 3312–3321, 2013.

Drake, S. I., M. E. Carunchia, M. A. Drake, P. Courtnet, K. Flinger, J. Jenkins & C. Pruitt. Sources of umami taste in cheddar and Swiss cheeses. *J. Food. Sci.* **72**, S360-S366, 2007.

Fernstrom, J. D. (ed.) 100th Anniversary Symposium of Umami Discovery: The Roles of

Glutamate in Taste, Gastrointestinal Function, Metabolism, and Physiology. *Amer. J. Clin. Nutr.* **90**, 705S-885S, 2009.

Fujii, M. *The Enlightened Kitchen: Fresh Vegetable Dishes From the Temples of Japan*. Kodanska Int., Tokyo, 2005.

Fuke, S. & Y. Ueda. Interactions between umami and other flavor characteristics. *Trends. Food. Sci. Technol.* **7**, 407–411, 1996.

Galindo, M. M., N. Voigt, J. Stein, J. van Lengerich, J.-D. Raguse, T. Hofmann, W. Meyerhof & M. Behrens. G-protein-coupled receptors in human fat taste perception. *Chem. Senses* **37**, 123–139, 2012.

Galindo-Cuspinera, V. & P. A. Breslin. The liason of sweet and savory. *Chem. Senses* **32**, 221–225, 2006.

Giacometti, T. Free and bound glutamate in natural products. *Glutamic Acid: Advances in Biochemistry and Physiology* (L. J. Filer, Jr. et al., eds.) Raven Press, New York, pp. 25–34, 1979.

Grocock, C. & S. Grainger. *Apicius*. Prospect Books, Devon, 2006.

Halpern, B. P. Glutamate and the flavor of foods. *J. Nutr.* **130**, 910S-914S, 2000.

Hoskin, R. *A Dictionary of Japanese food.* Tuttle Publ., Boston, 1996.

Ikeda, I. New seasonings. *Chem. Senses* **27**, 847–849, 2002. [Translated from *J. Chem. Soc. Japan* **30**, 820–836, 1909.]

Ismail, A. A. & K. Hansen. Accumulation of free amino acids during cheese ripening of some types of Danish cheese. *Milchwissenschaft* **27**, 556–559, 1972.

Jyotaki, M., N. Shigemura & Y. Ninomiya. Multiple umami receptors and their variants in human and mice. *J. Health. Sci.* **55**, 647–681, 2009.

Kasabian, A. & D. Kasabian. *The Fifth Taste: Cooking with Umami*. Universe Publishing, New York, 2005.

Kawai, M., H. Uneyama & H. Miyano. Taste-active compounds in foods, with concentration on umami compounds. *J. Health Sci.* **55**, 667–673, 2009.

Kawamura, Y. & M. Kare. *Umami: A Basic Taste: Physiology, Biochemistry, Nutrition, Food Science*. Marcel Dekker Inc., New York, 1986.

Komata, Y. Umami taste of seafoods. *Food Rev. Int.* **6**, 457–487, 1990.

Krebs, J. R. The gourmet ape: evolution and human food preferences. *Am. J. Clin. Nutr.* **90**, 707S-711S, 2009.

Kremer, S., J. Mojet & R. Shimojo. Salt reduction in foods using naturally brewed soy sauce. *J. Food Sci.* **74**, S255-S262, 2009.

Kuninaka, A. Studies on taste of ribonucleic acid derivatives. *J. Agric. Chem. Soc. Jpn.* **34**, 487–492, 1960.

Kwok, R. H. M. Chinese restaurant syndrome. *New Eng. J. Med.* **278**, 796, 1968.

Lanfer, A., K. Bammann, K. Knof, K. Buchecker, P. Russz, T. Veidebaum, Y. Kourides, S. de Henauw, D. Molnar, S. Bel-Serrat, L. Lissner & W. Ahrens. Predictors and correlates of taste preferences in European children: The IDEFICS study. *Food Quality and Preference* **27**, 128–136, 2013.

Li, X., L. Staszewski, H. Xu, K. Durick, M. Zoller & E. Adler. Human receptors for sweet and umami taste. *Proc. Natl. Acad. Sci. USA* **99**, 4692–4696, 2002.

Löliger, J. Function and importance of glutamate for savory foods. *Amer. Soc. Nutr. Sci.* **130**, 915S-920S, 2000.

Maga, J. A. Flavor potentiators. *Crit. Rev. Food Sci. Nutr.* **18**, 231–312, 1983.

Marcus, J. B. Culinary applications of umami. *Food Technol.* **59**, 24–30, 2005.

Maruyama, Y., R. Yasyuda, M. Kuroda & Y. Eto. Kokumi substances, enhancers of basic tastes, induce responses in calcium-sensing receptor expressing taste cells. *PLOS ONE* **7**, e34489, 2012.

McGee, H. *On Food and Cooking: The Science and Lore of the Kitchen*. Scribner, New York, 2004.

Mouritsen, O. G. *Sushi: Food for the Eye, the Body & the Soul*. Springer, New York, 2009.

Mouritsen, O. G. Umami flavour as a means to regulate food intake and to improve nutrition and health. *Nutr. Health* **21**, 56–75, 2012.

Mouritsen, O. G. *Seaweeds: Edible, Available & Sustainable.* Chicago University Press, Chicago, 2013.

Mouritsen, O. G., L. Duelund, L. A. Bagatolli & H. Khandelia. The name of deliciousness and the gastrophysics behind it. *Flavour* **2**:9, 2013.

Mouritsen, O. G. & H. Khandelia. Molecular mechanism of the allosteric enhancement of the umami taste sensation. *FEBS J.* **279**, 3112–3120, 2012.

Mouritsen, O. G., L. Williams, R. Bjerregaard & L. Duelund. Seaweeds for umami flavour in the New Nordic Cuisine. *Flavour* **1**:4, 2012.

Murata, K., H. Ikehata & T. Miyamoto. Studies on the nutritional value of tempeh. *J. Food. Sci.* **32**, 580–586, 1967.

Nakajima, N., K. Ishikawa, K. Kamada & E. Fujita. Food chemical studies on 5'-ribonucleotides. Part I. On the 5'-ribonucleotides in foods (1) Determination of the 5'-ribonucleotides in various stocks by ion exchange chromatography. *J. Agric. Chem. Soc. Jpn.* **35**, 797–803, 1961.

Nature Outlook. Making sense of flavour. *Nature* **486**, 21 June 2012, S1-S48, 2012.

Nelson, G., J. Chandrashekar, M. A. Moon, L. Feng, G. Zhao, N. J. Ryba & C. S. Zuker. An amino acid taste receptor. *Nature* **416**, 199–202, 2002.

Ninomiya, K. Natural occurrence. *Food. Rev. Int.* **14**, 177–211, 1998.

Ninomiya, K. Umami: A universal taste. *Food Rev. Int.* **18**, 23–38, 2002.

O'Mahony, M. & R. Ishii. A comparison of English and Japanese taste languages: Taste descriptive methodology, codability and the umami taste. *British J. Psychol.* **77**, 161–174, 1986.

Oruna-Concha, M.-J., L. Methven, H. Blumenthal, C. Young & D. S. Mottram. Differences in glutamic acid and 5'-ribonucleotide contents between flesh and pulp of tomatoes and the relationship with umami taste. *J. Agric. Food Chem.* **55**, 5776–5780, 2007.

Osawa, Y. Glutamate perception, soup stock, and the concept of umami: the ethnography, food ecology, and history of dashi in Japan. *Ecol. Food Nutr.* **51**, 329–355, 2012.

Otsuka, S. Umami in Japan, Korea, and Southeast Asia. *Food. Rev. Int.* **14**, 247–256, 1998.

Prescott, J. *Taste Matters: Why We Like the Food We Do.* Reaktion Books, London, 2012.

Rhatigan, P. *The Irish Seaweed Kitchen.* Booklink, Co Down, Ireland, 2010.

Roberts, M. *Secret Ingredients: The Magical Process of Combining Flavors.* Bantam Books, New York, 1988.

San Gabriel, A. M., T. Maekawa, H. Uneyama, S. Yoshie & K. Torii. mGluR1 in the fundic glands of rat stomach. *FEBS Lett.* **581**, 1119–1123, 2007.

San Gabriel, A. & H. Uneyama. Amino acid sensing in the gastrointestinal tract. *Amino Acids* **45**, 451–461, 2013.

Schiffman, S. S. Sensory enhancement of foods for the elderly with monosodium glutamate and flavors. *Food Rev. Int.* **14**, 321–333, 1998.

Shepherd, G. *Neurogastronomy.* Columbia University Press, New York, 2011.

Shurtleff, W. & A. Aoyagi. *The Book of Miso.* Ten Speed Press, Berkeley, California, 2001.

Soldo, T., I. Blank & T. Hofmann. (1) (S) Alapyridaine—a general taste enhancer? *Chem. Senses* **28**, 371–379, 2003.

Steiman, H. *Essentials of Wine: A Guide to the Basics.* Running Press, Philadelphia, 2000.

Stevenson, R. J. *The Psychology of Flavour.* Oxford University Press, Oxford, 2009.

Temussi, P. A. Sweet, bitter and umami receptors: a complex relationship. *Trends Biochem. Sci.* **34**, 296–302, 2009.

This, H. *Molecular Gastronomy.* Columbia University Press, New York, 2002.

This, H. *Kitchen Mysteries: Revealing the Science of Cooking.* Columbia University Press, New York, 2007.

Tsuji, A. *Japanese Cooking: A Simple Art*. Kodansha Intl., Tokyo, 1980.

Ueda, Y., M. Sakaguchi, K. Hirayada, R. Miyajima & A. Kimizuka. Characteristic flavor constituents in water extracts of garlic. *Agric. Biol. Chem.* **54**, 163–169, 1990.

Uneyama, H., M. Kawai, Y. Sekine-Hayakawa & K. Torii. Contribution of umami substances in human salivation during meal. *J. Med. Invest.* **56** Supplement, 197–204, 2009.

Ventura, A. K., A. San Gabriel, M. Hirota & J. A. Mennella. Free amino acid content in infant formulas. *Nutr. Food Sci.* **42**, 271–278, 2012.

Watanabe, K., H. L. Lan, K. Yamaguchi & S. Konosu. Role of extractive components of scallop in its characteristic taste development. *J. Jpn. Soc. Food Sci. Technol.* **37**, 439–445, 1990.

Williams, A. N. & K. M. Woessner. Monosodium glutamate 'allergy': menace or myth? *Clin. Exp. Allergy* **39**, 640–646, 2009.

Winkel, C., A. de Klerk, J. Visser, E. de Rijke, J. Bakker, T. Koenig & H. Renes. New developments in umami (enhancing) molecules. *Chem. Biodiver.* **5**, 1195–1203, 2008.

Wrangham, R. *Catching Fire*. Basic Books, New York, 2009.

Wrangham, R. W., J. H. Jones, G. Laden, D. Pilbeam & N. Conklin-Brittain. The raw and the stolen. *Curr. Anthropol.* **40**, 567–577, 1999.

Yasuo, T., Y. Kusuhara, K. Yasumatsu & Y. Ninomiya. Multiple receptor systems for glutamate detection in the taste organ. *Biol. Pharm. Bull.* **31**, 1833–1837, 2008.

Yamaguchi, S. & K. Ninomiya. What is umami? *Food Rev. Int.* **14**, 123–138, 1998.

Yamaguchi, S. & K. Ninomiya. Umami and food palatability. *J. Nutr.* **130**, 921S–926S, 2000.

Yoshida, Y. Umami taste and traditional seasonings. *Food Rev. Int.* **14**, 213–246, 1998.

Zhang, F. B., B. Klebansky, R. M. Fine, H. Xu, A. Pronin, H. Liu, C. Tachdjian & X. Li. Molecular mechanism for the umami taste synergism. *Proc. Natl. Acad. Sci. USA* **105**, 20930–20934, 2008.

Illustration credits

Jonas Drotner Mouritsen has taken the photographs on pp. xv, 21, 30, 37, 47, 53, 57, 59, 62, 63, 67, 68, 71, 75, 77, 78, 83, 84, 85, 86, 89, 90, 91, 96 (*top*), 98, 99, 107, 109, 110, 113, 114, 117, 118, 119, 121, 122, 127, 129, 131, 133, 134, 135, 137, 147, 149, 150, 153, 163, 164, 165, 171, 172, 174, 176, 177, 182, 184, 187, 189, 193, 195, 196, 198, 202, and 205, 215, 232, 238, and 254.

Jonas Drotner Mouritsen has drawn the illustrations on pp. 7, 16, 17, 29, and 42 (data from Yamaguchi. The synergistic taste effect of monosodium glutamate and disodium 5'-inosinate. *J. Food Sci.* **32**, 473–478, 1967).

The image of dried and cured Spanish hams on p. 9 is kindly provided by ICEX-Madrid (photographer Desconocido/ICEX).

The Archives of Ajinomoto Co., Inc. have permitted reproduction of the portrait of Kikunae Ikeda on p. 23 and the image of early industrial glutamate production on p. 28.

Morten Kringelbach has kindly given permission to use the images of brain scans on p. 43 (adapted from Araujo et al. Representation of umami taste in the human brain. *J. Neurophysiol.* **90**, 313–319, 2003).

Ole G. Mouritsen has taken the photographs on pp. 49, 73, 93, 94, 125, and 145.

The image of cheese maturation on p. 61 is kindly provided by CEX-Madrid (photographer Luis Carré/ICEX).

The image of the konbu harvesters in Hokkaido on p. 66 is kindly provided by Norishige Yotsukura.

Rien Bongers has given permission to reproduce the image of the Roman *garum* installation from Almuñécar in Spain on p. 79.

The image of *katsuobushi* production in Japan on p. 95 is from the Tokyo Foundation Website, http://www.tokyofoundation.org/en/topics/japanese-traditional-foods/vol.-15-dried-bonito. ©Kazuo Kikuchi. Reproduced by permission.

The photograph of *surströmming* on p. 96 (*bottom*) is taken by Wrote and is licensed under Creative Commons.

The photograph of *hákarl* from Drangsnes on Iceland on p. 97 is taken by Fulvio and Ida Attisani and is licensed under Creative Commons.

The engraving of the pike angler on p. 100 by Boyd Hanna is from Izaak Walton's book *The Compleat Angler*, The Peter Pauper Press, Mount Vernon, 1947.

The photographs of the pike dinner on pp. 102–103 are courtesy of Kristoff Styrbæk.

The graphic illustration of the contents of the various free amino acids in dashi and soups on p. 161 is drawn by Jonas Drotner Mouritsen and based on data from http://www.umamiinfo.com.

Jonas Drotner Mouritsen has drawn the illustrations on pp. 218–220.

The image of sun-ripened tomaotes on p. 212 is kindly provided by CEX-Madrid (photographer Amador Toril/ICEX).

Aktuelt Naturvidenskab has kindly given permission to use the graphic illustration of the Pac-Man analogy of the umami receptor on p. 221.

Himanshu Khandelia has produced the graphics of the molecular structures on pp. 221, 222, and 231.

◄ Classical Japanese woodblock print showing how bonito (*katsuo*) is caught with poles and lines. Utagawa (Ando) Hiroshige (1797-1858).

Glossary

acetic acid organic acid formed by bacterial and fungal fermentation of sugars. Vinegar contains diluted acetic acid.

adenosine triphosphate (adenosine-5'-triphosphate, ATP) polynucleotide that is the biochemical source of energy production in living cells. Among other substances, it can be transformed into the 5'-ribonucleotides inosinate, adenylate, and guanylate, which are associated with synergistic umami.

adenylate adenosine-5'-monophosphate (AMP), a salt of the nucleic acid adenylic acid; synergizes with glutamate to enhance umami; found especially in fish, shellfish, squid, and tomatoes.

adenylic acid nucleic acid; its salts, adenylates, are a source of umami synergy.

Ajinomoto international Japanese company that was founded in 1908 by the Japanese chemist who identified umami, Kikunae Ikeda, together with the entrepreneur Saburosuke Suzuki. Its primary product is MSG, which is used worldwide as a taste enhancer.

alanine amino acid with a sweetish taste.

alapyridaine tasteless chemical substance found in beef stock; enhances umami.

alkaloid one of a group of basic chemical compounds that are rich in nitrogen. The bitter-tasting substances caffeine and quinine are alkaloids.

allec type of fermented fish paste; made from the dregs left over from the production of the classic fish sauce *garum*.

allostery biochemical expression for the condition in which a substance can regulate the function of a protein by binding to it on a site different from the main active site. Simultaneous binding of glutamate and 5'-ribonucleotides to the umami receptor and the resulting synergy in stimulating the receptor is an example of allostery.

amino acid small molecule made up of between ten and forty atoms; in addition to carbon, hydrogen, and oxygen, it always contains an amino group. Amino acids are the fundamental building blocks of proteins. Examples include glycine, glutamic acid, alanine, proline, and arginine. Nature makes use of twenty different specific amino acids to construct proteins, which are chains of amino acids bound together with peptide bonds. Short chains are called polypeptides and long ones proteins. In food, amino acids are often found bound together in proteins and also as free amino acids that can have an effect on taste. An example is glutamic acid, which is the basis of umami. Of the twenty natural amino acids, there are nine, known as the essential amino acids, that cannot be produced by the human body and that we must therefore obtain from our food (valine, leucine, lysine, histidine, isoleucine, methionine, phenylalanine, threonine, and tryptophan). Amino acids are chiral molecules, meaning that they are found in two versions that are chemically identical but are mirror images of each other. They are referred to as left-turning (L-amino acids) and right-turning (D-amino acids). Their tastes can vary depending on which way they turn.

AMP see adenylate.

ankimo Japanese expression for liver from monkfish (*ankō*, *Lophius piscatorius*).

ao-nori species of green alga that is similar to sea lettuce.

Apicius, Marcus Gavius legendary Roman gourmet who lived in the first century CE, to whom authorship of the comprehensive ancient work on the culinary arts *De re coquinaria* is popularly attributed. The extensive use of an umami-rich ingredient, *garum* (q.v.), in this collection of recipes is evidence of an early intuitive awareness of the fifth taste in Western cuisine.

arabushi see *katsuobushi*.

arginine amino acid with a bitter taste.

aspartic acid amino acid; its salt, aspartate, imparts a little umami, about 8 percent as much as glutamate.

aspartame artificial sweetener, not based on sugars (saccharides). It is a dipeptide made up of two amino acids, namely, aspartic acid and phenylalanine, and is 150–200 times as sweet as sugar.

aspartate a salt of the amino acid aspartic acid; for example, monosodium aspartate (MSA), which imparts a small amount of umami. Large quantities of free aspartate are found in konbu and extracted into dashi.

Aspergillus glaucus mold species; used for the fermentation of *katsuobushi*.

Aspergillus oryzae mold species; an ingredient in the fermentation medium *kōji*, used for the production of soy sauce (*shōyu*), miso, mirin, and sake.

Aspergillus sojae mold species; used in the fermentation of soy sauce (*shōyu*).

astringency a mechanical sensory impression that is possibly linked to the taste cells in the mouth; commonly associated with the taste of tea or red wine. Both contain tannins, which react with proteins in the mucus on the surface of the tongue and saliva, causing sensations of sharpness, dryness, and friction. Depending on context, astringency can be registered as pleasant or unpleasant.

ATP see adenosine triphosphate.

autolysis special case of hydrolysis, also known as self-digestion, in which the cell is broken down by its own enzymes. An example is the hydrolysis of yeast to make yeast extract.

azuki small red or green beans (*Phaseolus angularis*). The red variety tastes sweet and is used in the form of a paste in Japanese cakes, desserts, and confections.

bacalao Spanish term for salted and dried cod that is a traditional food item in Galicia in northern Spain and in Portugal.

Bacillus subtilis nattō bacterial culture used to ferment soybeans to make *nattō*.

bagna càuda regional specialty from Piedmont, Italy, eaten as a sort of fondue or warm dip; made with olive oil, garlic, cream or milk, and many anchovy fillets.

bakasang fermented fish sauce from Indonesia.

Bayonne ham salted, air-dried French country ham, a product of the region around Bayonne in southern France.

biga Italian expression for a dry, yeasty dough (not sourdough) used in many kinds of Italian breads; for example, pizza dough and *ciabatta*.

biltong dried, vinegar-marinated, and salted meat, originally from South Africa, made, for example, from beef or emu. Often prepared with a variety of spices. Similar to jerky, but cut into thicker slices.

bisque shellfish soup with wine, vegetables, and herbs, thickened with cream.

Bloody Mary vodka cocktail that is rich in umami due to tomato juice and Worcestershire sauce.

Bolognese sauce Italian meat sauce with tomato; used on pasta.

botargo (*bottarga*) dried fish roe from tuna, cod, or gray mullet; considered a major delicacy in Spain (*botargo*) and Italy (*bottarga*).

bottarga see *botargo*.

bouillabaisse thick fish and shellfish soup, often with vegetables and eggs; originally from Provence in France.

bouillon (Fr. *bouilli*: cooked) clear soup or stock, based, for example, on meat, poultry, or fish.

Bovril beef extract, popular in England, developed in the 1870s; now also available as a chicken or a yeast extract; widely used in soups and gravies or to make a simple hot broth.

braising method of cooking meat, often the less

tender cuts. Meat is first seared and then slow-cooked with only a little liquid; the combined effect results in more umami and a richer taste.

brandade dried salt cod that has been soaked in water, then blended with olive oil, garlic, cooked potatoes, or softened bread crumbs. It is a specialty of southern France and other areas around the Mediterranean and was originally eaten as Lenten food.

Brillat-Savarin, Jean Anthelme (1755–1826) French lawyer, politician, and judge, whose most famous work, *The Physiology of Taste*, was published in 1825 and has not been out of print since. He is considered one of the fathers of gastronomy.

brioche light bread made with a yeast dough that is rich in butter and eggs; can be served with a savory stuffing or as a sweet dessert.

brunoise mixture of finely diced braised vegetables used as a foundation for stuffing, sauces, and soups.

budu Malaysian fermented fish sauce.

caffeine bitter, naturally occurring substance (alkaloid), found in coffee, tea, and some other plants.

capsaicin organic substance responsible for the strong, astringent taste of chile peppers.

Carême, Marie-Antoine (1783–1833) French chef and writer, considered the greatest exponent of haute cuisine.

carpaccio originally very thin slices of raw beef and veal. The term is now also used for other very thin slices of foods such as fish.

casein protein in milk.

cassoulet slow-cooked casserole originating in Languedoc made from white beans and various meats, such as sausage and duck.

catalyst substance that modifies or accelerates a chemical reaction without itself being consumed in the process. Enzymes function as catalysts in biochemical reactions; for example, in the breakdown of proteins to free amino acids.

cell membrane the thin layer of material that surrounds a cell; made up of fats, proteins, and carbohydrates.

Cenovis Swiss-made yeast extract spread that resembles Marmite and Vegemite.

chemesthesis technical term used in sensory science to describe the sensitivity of the skin and mucus membranes to the chemical stimulation of the nerve endings of the trigeminal nerve. This can lead to irritation, which may be an indication that the stimulus may be harmful. An example of chemesthesis is the painful sensation on the tongue caused by various substances in chile pepper (capsaicin), black pepper (piperine), and mustard (isothiocyanate), which is associated with a piquant taste. The special tingling in the mouth and nose from carbonated drinks is another example. Sensations of warmth and cold, which are also registered in the nerve endings in the mouth, are related to chemesthesis.

'Chinese restaurant syndrome' a complex, vaguely defined condition characterized by symptoms such as a flushing, sensitivity in the throat, headache, general weakness, heart palpitations, and tightness of the chest (MSG symptom complex). Some have attributed this to the ingestion of MSG. The best well-documented clinical investigations have, however, been unable to demonstrate any negative effects of MSG.

chorizo air-dried, fermented Spanish sausage.

confit a prepared food that is cooked and preserved in a medium, usually oil or fat. Examples include *confit* of salted goose or duck.

corned beef beef that has been pickled in a spiced brine and then cooked; *corn* refers to the coarse grains of salt used in the process.

cranial nerves nerves that emerge directly from the cerebrum or the brain stem. Some cranial nerves ensure that sensory perceptions are registered in the brain. The trigeminal nerve is a cranial nerve that communicates mouthfeel to the brain; it is independent from the facial nerves that convey taste impressions. The vagus nerve is a cranial nerve that connects some parts of the stomach with the brain.

crouton very small piece of bread that has been toasted in butter or oil.

daikon Japanese expression for a large white radish (*Raphanus sativus*), also called Chinese radish.

dashi Japanese expression meaning cooked extract; a stock made from, for example, seaweed (konbu) and bonito fish flakes (*katsuobushi*). First dashi (*ichiban* dashi) and second dashi (*niban* dashi) refer to the first and second extract, respectively. Konbu dashi is based solely on the seaweed; *niboshi* dashi makes use of small dried fish, *niboshi*, instead of *katsuobushi*. *Shōjin* dashi is a purely vegetarian stock in which shiitake mushrooms are substituted for *katsuobushi*. Dashi powder (Hon-dashi) is dehydrated dashi, used as a soup powder.

deoxyribonucleic acid see DNA.

DNA deoxyribonucleic acid; polynucleotide consisting of a long chain of nucleotides (q.v.) formed from the sugar group deoxyribose and the bases adenine, guanine, thymine, and cytosine. In DNA there are normally two polynucleotide chains which spiral around each other in a double helix. DNA is the basis for the genetic information encoded in the genome and in genetic material.

donko highly prized shiitake mushroom with a small, dark cap; in dried form it contains a great deal of umami.

douchi Chinese expression for fermented, salted black soybeans used to make seasoning sauces and pastes.

dulse the red alga *Palmaria palmata*.

enokitake winter mushroom (*Flammulina velutipes*) that grows in a cluster; has long, thin white stalks and a small cap.

enzyme protein that functions as a catalyst in a chemical or a biochemical reaction. Proteolytic enzymes can split proteins into small peptides or free amino acids.

estofado Spanish word for a casserole dish or stew made, for example, with beef.

fermentation process in which microorganisms (yeast or bacteria) or enzymes convert carbohydrates into alcohols or acids. For example, yeast cells convert sugars to alcohol or to vinegar.

fish sauce sauce made from fish or shellfish, usually by salting and fermenting them either whole or else the parts that would otherwise be discarded, such as blood and innards. Some are made with fresh fish, others with ones that have been dried first; anchovies are included in most of the sauces. Fermentation takes place using the enzymes found in the animal products, releasing an abundance of free amino acids, especially alanine and glutamate. *Garum* made in ancient Greece and Rome is the oldest example of a fish sauce in Western cuisine, and a whole range of fish sauces are commonly used in Asian cuisine.

flavor collective term for the sensory impression made by a food, both its taste and smell (its aroma substances), as well as its mouthfeel and chemesthesis.

folic acid vitamin B_9.

fricassee stew usually made with meat and vegetables in a light sauce, such as veal or poultry in a white sauce or lamb with an egg and lemon sauce.

fu Japanese expression for kneaded wheat gluten, also known in Chinese as *seitan*; in raw form, it is *nama fu*, and roasted or dried, *yaki fu*.

fundic glands glands located in the stomach that secrete certain enzymes (proteases) that break down proteins.

furu (or *sufu*) Chinese expression for fermented tofu. If it has been fermented for a long time, it is also called 'stinky' tofu.

fushi Japanese expression for fish that is preserved by cooking, drying, salting, smoking, and fermenting. *Katsuobushi* and *magurobushi* are types of *fushi*.

ganjang Korean soy sauce.

garon (*garos* or *garus*) see *garum*.

garum (*garon, liquamen*) brownish liquid that seeps out when salted small fish and fish innards of, for example, mackerel and tuna are crushed and fermented for a long time. *Garum* production was an important industry in Rome and Greece in antiquity. *Oxygarum* is *garum* mixed with wine vinegar and *meligarum* with honey. *Garum* is similar to East Asian fish sauces.

glucose sugar; a monosaccharide that is the most important carbohydrate in plants and animals. In plants and algae, it is formed by photosynthesis.

glutamate salt of the amino acid glutamic acid; for example, monosodium glutamate (MSG, $C_5H_8NO_4Na$). In water, glutamate splits into sodium ions and glutamate ions, the latter being the main source of umami.

glutamic acid amino acid. It was identified in 1866 by the German chemist Karl Heinrich Leopold Ritthausen, who was studying proteins in wheat, but he seems to have taken little interest in his discovery. In the early 1900s, the chemist Emil Fischer studied different amino acids and noted that glutamic acid has an insipid and slightly sweet taste. The salts from glutamic acid are called glutamates (q.v.).

glutathione tripeptide that can be synthesized in the body from the three amino acids cysteine, glutamic acid, and glycine. Can act as an antioxidant and is one of the peptide compounds that stimulate the *kokumi* taste.

gluten certain proteins (especially gliadin and glutenin) found in wheat. *Fu* and *seitan* are concentrated gluten.

glycine amino acid with a sweetish taste.

GMP see guanylate.

G-protein-coupled receptor transmembrane protein with seven segments that pass through the cell membrane. It also has a large extracellular part that sticks out of and away from the cell membrane; on taste receptors it captures and identifies taste molecules.

guanylate guanosine-5'-monophosphate (GMP), guanylic acid salt; interacts synergistically with glutamate to enhance umami. Disodium, dipotassium, and calcium guanylates are used as food additives.

guanylic acid nucleic acid discovered in 1898 by the Norwegian physician Ivar Bang, who was studying the nucleic acids in the pancreas.

gunkan-maki sushi made by enclosing rice with a piece of nori and placing the fill on top (battleship sushi).

gustducin G-protein that can bind to the taste receptors T1R and T2R.

gyokuro Japanese green tea of the very finest quality.

Hidaka-konbu see konbu.

hishio forerunner of Japanese soy sauce (*shōyu*), known as early as the eighth century; made from fermented soybeans to which rice, salt, and sake were added.

histidine amino acid with a bitter taste.

hōjicha roasted green tea.

homeostasis the equilibrium achieved when processes with opposite effects work together in a body to achieve a form of internal self-regulation and stability.

HP sauce thick brown sauce made from vinegar and spices; used to enhance umami.

hydrolysis (Gr. *hydro*: water, *lysis*: splitting) chemical process by which a molecule is split into smaller parts by absorption of water. An example is hydrolysis of animal or vegetable protein to release a free amino acid, such as glutamate.

Autolysis is a special case of hydrolysis, which takes place without any external means; for example, the hydrolysis of yeast proteins to produce yeast extract with the help of the yeast's own enzymes.

hydrolyzed protein food product containing amino acids; for example, glutamate, made by hydrolyzing proteins from vegetables, animals, or fungi. It is not considered to be a food additive and is used as a taste enhancer to add umami; for example, in soup powders. Yeast extract is an example of hydrolyzed protein.

Ikeda, Kikunae (1864–1936) Japanese chemist who coined the word umami and was the first to use chemical methods to investigate it. Demonstrated that monosodium glutamate (MSG) is the substance in seaweeds (specifically, konbu) and, therefore, in dashi that imparts umami.

ikijime a 350-year-old Japanese technique for killing fish, which is intended to ensure that the fish has the best taste and suffers the least damage to, and discoloration of, the flesh. It works by delaying the onset of *rigor mortis*. The fish dies without being stressed, which releases more of the substances (especially inosinate) that are sources of umami.

IMP see inosinate.

inosinate inosine-5'-monophosphate (IMP), inosinic acid salt, commonly found in meat, fish, and shellfish; interacts synergistically with glutamate to enhance umami. Disodium, dipotassium, and calcium inosinates are used as food additives.

inosinic acid nucleic acid identified in 1847 by the German chemist Justus von Liebig (1803–1873), who isolated the acid from beef soup stock. Inosinic acid salts, inosinates, contribute to synergistic umami.

ion channel protein located in the cell membrane that functions as a channel that can, selectively, allow ions to pass through. Ion channels in the neural cells, for example, sensory cells, are involved in the generation of electrical signals that send information to the brain about a sensory impression, such as one due to a taste substance.

irori traditional Japanese fish sauce, *katsuo-irori*, made as a kind of paste by reducing the water in which bonito (*katsuo*) has been cooked. *Irori* is no longer made, but a similar product, *senji*, is still available on Kyushu, an island in the southern part of Japan.

irritant chemical substance that affects the nerve endings of the trigeminal nerve (chemesthesis). This leads to an irritation, which, in principle, is a warning signal indicating that mucus cells are in danger of being damaged. A well-known example is astringency.

ishiri (also known as *ishiru*) Japanese fermented fish sauce with a high glutamate content.

isothiocyanate chemical compound with the S=C=N– group. It has an pungent odor, which is formed, for example, when mustard seeds, cabbage, horseradish, or *wasabi* are crushed, and has an astringent taste.

jamón serrano (*jamón iberico de bellota*) air-dried mountain ham; Spanish and Portuguese specialty made from the meat of Landrace and wild pigs.

jerky dried, salted, and smoked meat; similar to *biltong*.

jiàng Chinese miso. *Jiàng yóu* is Chinese soy sauce. *Majiàng* is a type of fish paste with soybeans.

kabayaki Japanese expression for a dish with fish or shellfish, typically eel, which is filleted and dipped in a sweet soy-based sauce, such as teriyaki sauce, before being broiled.

kaiseki (*cha-kaiseki*) the formal meal served at a traditional Japanese tea ceremony.

karebushi see *katsuobushi*.

katsuo Japanese fish, bonito or skipjack tuna (*Katsuwonus pelamis*), related to mackerel and tuna;

it is widely used to make *katsuobushi* (*arabushi* and *karebushi*).

katsuobushi Japanese term for a hard fillet of *katsuo* (bonito) that has undergone an extensive process involving cooking, drying, salting, smoking, and sometimes fermenting. *Katsuobushi* is the ideal exemplar of an ingredient that has great quantities of inosinate, which interacts synergistically to elicit umami. There are two main types of *katsuobushi*: *arabushi*, which is not fermented and has a milder taste, and *karebushi*, which is fermented and harder. There are also two types of *karebushi*: On one the red side of the fillet is kept and on the other (*chinuki katsuobushi*) it is cut off. The second type has a milder, less bitter, and more delicate taste.

ketchup tomato purée made from sun-ripened tomatoes, to which may be added mushrooms, anchovies, vinegar, walnuts, pickled vegetables, and a number of spices.

kimchi fermented cabbage, considered a national dish in Korea.

Knorr, Carl Heinrich Theodor (1800–1875) German food products manufacturer. In the early 1870s, his company marketed the first commercially prepared soup powders.

kobujime Japanese technique combining fresh fish with konbu, e.g., for sashimi.

Kodama, Shintaro Japanese researcher who, in 1913, discovered that the substance in dried bonito (*katsuobushi*) that elicits umami is inosinate, the salt of a nucleic acid.

koe-chiap original Chinese expression for ketchup.

kōji Japanese expression for the fermentation medium made from cooked rice, soybeans, and wheat or barley to which is added a mold called *Aspergillus orzae*. *Kōji* is used in the production of soy sauce, miso, and sake; it contains a wide range of taste and aromatic substances, among them free amino acids and large quantities of glutamate.

kokumi taste or taste-enhancing effect that is claimed to be distinct from the five basic tastes. It combines three distinct elements: thickness—a rich complex interaction among the five basic tastes; continuity—the way in which long-lasting sensory effects grow over time or an increase in aftertaste; and mouthfeel—the reinforcement of a harmonious sensation throughout the whole mouth. *Kokumi* is evoked by the stimulus of certain calcium-sensitive channels on the tongue by small peptides such as glutathione and other gamma-glutamyl peptides, found in such foods as scallops, fish sauce, garlic, onions, and yeast extract. *Kokumi* substances have no taste of their own but can enhance saltiness, sweetness, and umami in addition to suppressing bitterness. The effect on sourness is still unclear.

kombucha fermented black tea.

konbu (kombu) species of large brown alga (*Saccharina japonica*) containing great quantities of glutamate; together with *katsuobushi* an essential ingredient in Japanese dashi. Of the many different variants of Japanese konbu, *ma-konbu*, *Rausu-konbu*, and *Rishiri-konbu* are considered to be the best bases for dashi and they yield a very light dashi with a mild and somewhat complex taste. A lower quality konbu is *Hidaka-konbu*. *Oboro-konbu* and *tororo-konbu* are dried konbu that have been marinated in rice vinegar and, after being dried partially, shaved or cut into paper-thin shavings. Konbu dashi is a water extract of konbu.

Kuninaka, Akira Japanese researcher who, in 1957, discovered that the 5′-ribonucleotide guanylate elicits umami. Subsequently a team of Japanese scientists found that dried shiitake mushrooms have large amounts of guanylate.

kuragakoi Japanese expression for the cellar aging of konbu to make it taste milder and increase its umami potential.

kusaya Japanese expression for horse mackerel or

flying fish that are cleaned and deboned, then brined and allowed to ferment in the warmth of the sun.

lactisole taste-modifying substance that suppresses the taste of certain natural and artificial sweeteners.

lactose sugar found in milk.

Laminariales taxonomic order that encompasses the large brown algae, for example, konbu (*Saccharina japonica*).

laver see *Porphyra*.

lecithin fat (phospholipid) found in all cell membranes; can be extracted, for instance, from egg yolks and soybeans. It acts as an emulsifier that can bind oil and water in emulsions such as mayonnaise.

lenthionine cyclic organic molecule containing carbon and sulfur; enzymatic formation of lenthionine is responsible for the distinctive aroma of shiitake mushrooms.

Lentinus edodes shiitake mushrooms.

leucine bitter-tasting amino acid.

lipid fat that consists of a water soluble part and an oil soluble part, which is normally a fatty acid. Biological membranes are composed of lipids.

liquamen Latin word for fermented fish sauce; see *garum*.

lysine sweet-tasting amino acid.

maccha Japanese powdered green tea.

Maggi, Julius (1846–1912) Swiss miller and industrial food pioneer who was the first to introduce concentrated umami, in the form of soup powders, to the European market.

magurobushi dried and fermented tuna, produced in the same manner as *katsuobushi*.

Maillard reactions class of chemical reactions that are typically associated with nonenzymatic browning occurring, for example, during frying, baking, or grilling. In the course of these reactions, carbohydrates bind with amino acids from proteins and, after a series of intermediate steps, form a number of poorly characterized brown pigments and aromatic substances collectively known as melanoids. These substances give rise to a broad spectrum of taste and smell sensations ranging from the flower- and plant-like to the meat- and earth-like.

ma-konbu see konbu.

Manganji **pepper** a Japanese pepper speciality from Kyoto; related to American bell pepper.

mannitol sugar alcohol found, for example, in large brown algae, among them, konbu and sugar kelp; it imparts a characteristic sweet taste.

Marmite trade name for a yeast extract that is a sticky, dark brown paste with an extremely salty taste. Contains large quantities of glutamate and vitamin B, especially folic acid (B_9) and B_{12}; used as a taste enhancer to bring out umami. The name is based on the French word for a large, covered earthenware cooking pot.

matsutake (*Tricholoma matsu*) edible pine mushroom; much sought after in Japan.

McGee, Harold American author who writes about the chemistry of food and cooking; his works include *On Food and Cooking*: *The Science and Lore of the Kitchen*.

medisterpølse traditional Danish sausage made from ground pork, pork fat, and spices; normally cooked in water and then fried.

meligarum see *garum*.

metabolism the physical and chemical processes in an organism that are required to maintain life; for example, the digestion of foodstuffs.

metallic taste not a true taste; actually the taste of oxidized products that are often formed when fats (lipids) come in contact with metals.

methionine amino acid with a bitter, sulfurous taste.

mGluR1 glutamate receptor in the brain.

mGluR4 metabotropic glutamate receptor in the

brain. It is a G-protein-coupled receptor, which is not itself an ion channel but whose action is indirectly coupled to an ion channel in the cell membrane.

Micrococcus glutamicus microorganism that grows in an anaerobic culture medium of nutrients and minerals; abundantly synthesizes and exudes glutamic acid onto its surface; the glutamic acid is released into the medium, from which it can be separated out and converted to MSG.

microvilli tiny hair-like protrusions extending from the membranes of some cells and organelles. In the case of the taste receptor cells, they are bundled together like the segments of an orange so that they form a pore on the surface of the tongue. Taste substances must pass through the pore in order to be detected. The receptors that biochemically identify the taste substances are located in the cell membranes.

mirin sweet rice wine with about 14 percent alcohol; made from steamed rice mixed with *kōji* to which rice brandy is added. Mirin is not intended to be drunk and is used as a taste additive to impart sweetness and umami.

miso Japanese fermented soybean paste made with *kōji*. Typically, miso paste contains 14 percent protein and large amounts of free amino acids, especially glutamate. The salt content varies from 5–15 percent. *Shiro* miso and *Shinshū* miso are white miso made with rice and soybeans, *aka* miso is red soybean miso, and *genmai* miso is made from brown rice. Some types of miso are made with barley (*mugi* miso), and a very dark, expensive variety is made with soybeans only (*Hatchō* miso). Vegetables pickled in miso are called *miso-zuke*, an example being *nasu-miso* made with eggplant. Miso soup is a soup made with dashi and miso.

mojama Spanish word for salted, air-dried tuna fillet.

molasses viscous, concentrated syrup that is a by-product of the refining of sugar cane or sugar beets.

monosodium aspartate see aspartate.

monosodium glutamate see glutamate.

moromi solid mass of fermented soybeans; one of the stages in the production of Japanese soy sauce.

MSA monosodium aspartate; the sodium salt of aspartic acid, which has a faint umami taste. See aspartate.

MSG monosodium glutamate; the sodium salt of glutamic acid; the most important source of umami. See glutamate.

mouthfeel collective term for sensory perceptions that are neither taste nor aroma but that interact very closely with them; influenced by the structure, texture, and morphological complexity of a food item. It is, to a very large extent, responsible for our overall impression of the food and can involve physical and mechanical impressions such as chewiness, viscosity, mouthcoating, and crunchiness.

muria Latin word for fermented fish sauce made with tuna.

mycelium branched filament that makes up the root mat of a fungus.

nabe (*nabemono*) Japanese fondue; vegetables, meat, mushrooms, and fish are cooked in a soup stock, such as dashi.

nam-pa Laotian expression for fermented fish sauce.

nam-pla Thai expression for fermented fish sauce.

nama-zushi see sushi.

nare-zushi see sushi.

nasu dengaku traditional Japanese dish made with deep-fried eggplant and miso.

nattō whole, small soybeans fermented to form a stringy, viscous mass with a distinctive flavor, a strong aroma, and an intense umami taste. Very protein-rich foodstuff with a good quantity of vitamin K. In contrast to many other fermented soybean products, it has only a little salt.

Glossary   247

neurotransmitter chemical substance that carries the signals between the neural cells. Glutamate is a neurotransmitter, which binds to the glutamate receptor in the membranes of the neural cells.

New Nordic Cuisine widespread, gastronomically inspired movement launched in 2004. It focuses on the development of a distinctive Nordic cuisine, built on traditional methods and the use of regional and seasonal raw ingredients of the highest quality, all with a view to enhancing deliciousness and promoting wellness.

ngan-pya-ye fermented fish sauce from Myanmar.

ngapi fermented shrimp paste from Myanmar.

niboshi cooked and sun-dried small fish, such as anchovies and sardines, or the young fish of larger fish, such as flying fish, some species of mackerel, and dorade. In some cases, the fish are also smoked or grilled (*yakiboshi*), which results in a stronger taste. *Niboshi* have both glutamate and inosinate, which interact synergistically to impart umami.

nimono Japanese expression for slow-cooked dishes.

nojime Japanese expression for fish that is aged before being eaten.

nori paper-thin sheets made from the blades of the red alga *Porphyra*, which are chopped, dried, pressed, and sometimes toasted. Among other uses, they are essential for making sushi rolls (*maki-zushi*).

nucleic acid chain of nucleotides (polynucleotide); for example, DNA and RNA.

nucleoside chemical compound consisting of a heterocyclic base bound to a sugar group. The base can be either a purine (adenine and guanine) or a pyrimidine (cytosine in RNA and DNA, as well as thymine in DNA and uracil in RNA). Nucleosides form part of DNA and RNA, where the sugar groups are deoxyribose and ribose, respectively.

nucleotide a nucleoside that has been phosphorylated; that is, it is made up of a base, a sugar group, and one or more phosphate groups. Phosphorylation typically takes place at the 5'-position, which is why these nucleotides are also called 5'-nucleotides. Nucleotides can polymerize to form a long chain, a polynucleotide; for example, DNA and RNA. IMP, GMP, AMP, and ATP are known as ribonucleotides (5'-ribonucleotides), as their sugar group is ribose.

nuoc mam tom cha Vietnamese fermented fish sauce.

nutritional yeast deactivated, dried baker's yeast or brewer's yeast (*Saccharomyces cerevisiae*) sold in the form of flakes. It is different from yeast extract and is an excellent source of vitamin B and umami.

oboro-konbu see konbu.

orthonasal refers to the sensory perception of odor substances that enter the nose from the external environment.

osmazôme old chemical expression coined by the French chemist Louis Jacques Thénard in 1806 (Gr. *osmē*: odor, *zomos*: soup) to describe the tasty bouillon resulting from boiling meat in water. Transliterated as osmazome in English.

osso buco dish made with beef or veal shanks braised in wine or broth, typically with vegetables, tomatoes, and herbs.

Ostwald, Friedrich Wilhelm (1853–1932) German chemist who was awarded the Nobel Prize in chemistry in 1909 for his work on chemical equilibria and catalysis. He is regarded as one of the founders of physical chemistry.

oxygarum see garum.

Palmaria palmata the red alga dulse.

pancetta Italian specialty; pork belly meat that is salted, spiced, and cured, then rolled up into a sausage shape; often used like bacon.

panko dried white Japanese bread crumbs.

Parmigiano-Reggiano the original Italian Parmesan cheese; hard, dry, aged cheese with a substantial glutamate content.

pata negra dried ham from black-foot pigs; a type of *jamón serrano*.

patis Philippine fermented fish sauce.

Penicillium roqueforti fungus used for the production of blue cheeses; for example, Roquefort, Stilton, Gorgonzola, and Danish blue.

pepperoni type of very thin, dried, spiced sausage.

peptide chemical compound made up of amino acids bound together in a chain with peptide bonds; long peptides are called polypeptides or proteins.

phenylalanine amino acid with a bitter taste.

pickling culinary technique of preserving food products, such as vegetables, fungus, fish, or eggs, by aging them in a vinegar solution or a brine.

piperine organic substance that imparts the strong taste in black pepper.

Porphyra (laver) red alga genus made up of about 70 distinct species; used for the production of nori.

presynaptic cell cell that plays a role in the transmission of nerve signals, for example, in the taste buds. In contrast to the taste receptor cells, the presynaptic cells can respond to several different types of tastes, given that they receive signals from several different taste receptor cells.

proline amino acid with a sweetish taste.

Promite trade name for a yeast extract similar to Marmite and Vegemite.

prosciutto *crudo* (Parma ham) air-dried ham, originally from central and northern Italy, made from pork. Prosciutto, like the Portuguese *presunto*, means something that has been dried thoroughly. Prosciutto *di cavallo* is made from horse meat.

protein polypeptide, that is, a long chain of amino acids bound together by peptide bonds. Receptors, which receive chemical signals and identify sensory impressions such as taste and smell, are proteins. Enzymes are a particular class of proteins, whose function is to ensure that chemical reactions take place under controlled circumstances. Proteins lose their functional ability (denature) and their physical properties change when they are heated, exposed to salt and acid (for example, when cooked, salted, or marinated), or degraded by enzymes during fermentation. When proteins are broken down, smaller peptides and free amino acids, for example, glutamic acid, are formed.

proteolytic enzyme (protease) enzyme that can break down proteins to peptides and free amino acids.

quinine natural, bitter-tasting substance (alkaloid) derived from the bark of the cinchona tree.

ragout stew made from meat, poultry, or game with vegetables and seasonings.

rakfisk Norwegian specialty made from salted fermented trout, whitefish, char, or, occasionally, perch that are caught in the early autumn.

ratatouille very filling dish made from vegetables, such as tomatoes, eggplant, onions, and celeriac, which are simmered in olive oil.

Rausu-konbu see konbu.

receptor protein molecule that has a special ability to recognize and bind to a particular substance; for example, a smell or taste molecule. Receptors are found in all membranes, especially those of nerve cells. A G-protein-coupled receptor is a receptor that binds a G-protein in a signaling process; for example, in connection with the transmission from a taste cell.

rennet liquid containing proteases (viz. enzymes) that can break down the milk protein casein into smaller peptides and free amino acids; used in cheese production because it causes milk to coagulate.

retronasal term used to designate the perception of odor substances that are released in the oral cavity and from there move up into the nose.

Rhizopas oligosporus fungus used for fermenting soybeans to make tempeh.

ribonucleotide see nucleotide.

rikakuru fermented fish extract from the Maldives.

Rishiri-konbu see konbu.

risotto dish made with special varieties of round medium- or short-grain rice that is sautéed in butter or oil and cooked in stock and/or wine, to which many other ingredients can be added. The rice absorbs the liquid and becomes soft, without losing the shape of the individual kernels.

RNA (ribonucleic acid) polynucleotide that, like DNA, is made up of four bases, but with uracil instead of thiamine and the sugar group ribose instead of deoxyribose.

rouille Provençal sauce made with olive oil, chile pepper, garlic, saffron, and, possibly, egg yolk.

Saccharina japonica Japanese species of large brown alga, konbu, which is the organism that contains the most free glutamate, typically 2.000–3.000 mg/100 g, depending on the variety.

saccharine artificial sweetener that is 300–400 times sweeter than ordinary sugar (sucrose) but that has a bitter, metallic aftertaste.

Saccharomyces cerevisiae baker's yeast.

sake rice wine made from polished rice that is cooked and then fermented with the help of a fermentation medium, *kōji*, which contains enzymes that can break starch down to sugar and proteins to free amino acids. A yeast culture converts the sugar to alcohol.

sake kasu lees left over from fermentation of sake.

salsa verde cold green sauce, usually made from parsley, capers, onions, garlic, anchovies, and olive oil; sometimes mustard is added.

salsiccia secca Italian air-dried sausage.

salt cod dried salted cod that can be used, among other dishes, to make *bacalao* or *brandade*.

sanshō type of Japanese pepper, similar to Chinese Sichuan pepper.

sashimi Japanese expression for specially prepared raw fish or shellfish that is sliced crosswise into small pieces.

sauté panfry something quickly at high heat with only a little fat.

sea lettuce type of green alga (*Ulva lactuca*).

seitan Chinese expression for *fu*, a concentrated solid of kneaded wheat gluten.

sencha high-quality Japanese green tea.

senji see *irori*.

sensory science science related to human sensory perceptions, especially with regard to flavor (taste, smell, texture, mouthfeel, and chemesthesis).

shichimi Japanese spice mixture with seven different types of tastes chosen from the list: *sanshō* pepper, white and black sesame seeds, toasted or dried ground chile pepper with mustard, dried ginger, *ao-nori* (type of green alga that resembles sea lettuce), dried *shiso* leaves, dried peel of citrus fruits, and hemp seeds. *Nama-shichimi* is a mixture of fresh herbs, not necessarily the same as those already listed.

shiitake the mushroom *Lentinus edodes*; in dried form it contains large quantities of guanylate, a source of synergistic umami.

shiokara Korean and Japanese expression for salted fish or mollusks fermented in their own viscera. A classical Japanese dish is *ika no shiokara*, whole small squid that are fermented using the enzymes in their entrails.

shōchū brandy distilled from sake.

shōjin **dashi** strictly vegetarian soup stock that is an extract of konbu and dried shiitake mushrooms. It can also be prepared with dried daikon and eaten, for example, with salted or dried tofu.

shōjin ryōri the very specialized vegetarian temple

cuisine, originally brought to Japan in the sixth century with the spread of Buddhism and refined by the Zen monks in the 1300s. *Shōjin* is a Buddhist expression meaning devotion to the esthetic striving for spiritual awakening.

shōyu Japanese soy sauce; if made according to the traditional method, fermentation takes place over a two-year period.

sodium aspartate see MSA.

sodium chloride NaCl, table salt.

sodium glutamate see MSG.

sorbet frozen dessert made with fruit juice and sugar but no cream or egg yolks.

soufflé fluffy, light dish made with beaten eggs; can be either savory or sweet.

soumbala West African food made with fermented *néré* (*Parkia biglobosa*) seeds; similar to miso paste.

soy sauce seasoning liquid made from cooked soybeans fermented in a saline solution.

stew dish made with a variety of ingredients, such as meat and vegetables, cut into uniform-size pieces. Ragout and fricassee are also types of stew.

stock a clear soup base or bouillon. A light stock is made from white meats and vegetables. A dark stock is based on bones, meat, and herbs that are browned first. The Japanese dashi is also a type of stock made from an extract of konbu and *katsuobushi*.

Stroganoff beef Stroganoff; dish made with thin slices of beef glacéed with a sauce of white wine, sour cream, and stock, to which sautéed onions and mushrooms are added.

succinic acid organic acid that can impart a taste reminiscent of umami; found, for example, in sake and shellfish.

sucrose ordinary sugar, a disaccharide made up of glucose and fructose (fruit sugar).

suimono clear soup made from the first dashi (*ichiban* dashi).

surströmming Swedish specialty made with small Baltic herring that are salted and fermented.

sushi Japanese specialty consisting of cooked vinegared rice, a variety of toppings (such as fish, shellfish, omelette, and vegetables) and, in some cases, seaweeds. *Nare-zushi* (fermented sushi) is fermented fish, the original form of sushi. *Nama-zushi* (raw sushi) is modern sushi, made with fresh and usually raw fish. *Maki-zushi* is rolled sushi, with a sheet of nori either on the outside or the inside. *Gunkan-zushi* is in the form of a little boat.

Tabasco sauce hot, spicy sauce made with Tabasco peppers, vinegar, and salt.

table salt ordinary cooking salt, NaCl.

tamari Japanese expression ('accumulated liquid') that originally referred to the liquid that seeped out of the fermented soybean mass used to make soy sauce and miso paste. At that point, tamari was a by-product from the production of miso. Tamari is now a product in its own right, like soy sauce, but made from soybeans only, with no wheat.

tamarind (*Tamarindus indica*) tropical tree whose pod-shaped fruit is used as a spice in Indian cuisine; also used in the production of Worcestershire sauce.

tang Chinese expression for both soup and soup stock, normally made with bones or chicken. *Di tang* is the expression that is closest to 'stock.' *Xiang tang* means 'delicious soup' and is probably the Chinese expression that most closely describes a soup or stock with a strong umami taste.

tannin (tannic acid) umbrella term for a variety of polyphenolic compounds, which are bitter taste substances; found in young red wine, black tea, and smoked products, among others.

taste physiologically based perception of taste substances, which can bind to particular taste receptors in the taste buds on the tongue. It is generally acknowledged that there are five basic tastes, sour, salt, sweet, bitter, and umami, which can be combined to make up all other tastes. See also flavor.

taste buds onion-shaped clusters of 50–150 taste receptor cells embedded in tiny protrusions located primarily on top of the tongue, but also distributed over the soft palate, pharynx, epiglottis, and the entrance to the esophagus.

taste enhancer (flavor enhancer, food additive) substance, which often has little or no taste itself, that can intensify another taste substance. Glutamate, inosinate, guanylate, and glutathione are typical taste enhancers. Yeast extract and hydrolyzed protein are used as taste enhancers but are considered foodstuffs.

taste-**mGluR4** glutamate receptor in the taste cells of the oral cavity. It is a truncated and much less sensitive form of the glutamate receptor mGluR4, which is found in the neural cells of the brain.

taste receptor cells sensory cells in the taste buds that contain the receptors that can identify taste substances.

tempeh fermented soybean product, originally from Java, that is very rich in vitamin B.

tempura Japanese expression for deep-fried fish, shellfish, or vegetables.

teriyaki Japanese expression for fish, meat, fowl, or vegetables that are grilled and basted with a marinade of soy sauce, mirin, sake, sugar, and, optionally, spices, leaving their surfaces with a lustrous sheen.

teuk trei fermented fish sauce from Cambodia.

texture in sensory physiological terms defined as the sensory and functional manifestation of structural, mechanical, and surface properties of a foodstuff.

theanine amino acid that makes up more than half of the free amino acid content of green tea leaves; imparts umami.

third spice monosodium glutamate (MSG).

This, Hervé French chemistry professor, who is considered one of the fathers of molecular gastronomy.

thrush small songbird that is caught for food in some areas around the Mediterranean.

tofu coagulated, protein-rich solid made from soy milk. 'Stinky' tofu is fermented tofu.

tororo-konbu see konbu.

trigeminal nerve see cranial nerves.

trimethylamine foul-smelling organic substance (tertiary amine) produced, for example, by bacterial decomposition of trimethylaminoxide in dead seaweeds and fish.

tsuyu mixture of mirin, soy sauce, and sake to make a sauce that is rich in umami; used for dipping tempura and noodles.

T1R family of G-protein-coupled taste receptors (for example, T1R1, T1R2, and T1R3) that are activated by substances that have sweet and umami tastes.

T2R (also known as TAS2R) family of G-protein-coupled taste receptors that are activated by bitter taste substances.

umami Japanese expression for 'the fifth taste'; coined by the Japanese chemist Kikunae Ikeda in 1909 in connection with the identification of glutamate in dashi.

vagus nerve see also cranial nerves. The vagus nerve can send signals to the brain that the fundic glands in the stomach need to secrete certain enzymes (proteases) to break down proteins.

vegan vegetarian who neither eats nor uses any products derived from animals.

Vegemite trade name for an Australian yeast extract similar to Marmite; contains large quantities of free glutamate and, consequently, has umami.

vitamin one of a group of different essential

organic substances that the body itself can produce only in very limited quantities and that are, therefore, primarily derived from the diet. Examples are vitamins A, B, C, D, E, and K. Yeast extract contains very large quantities of vitamin B, especially folic acid (vitamin B_9) and vitamin B_{12}.

von Liebig, Justus (1803–1873) German chemist who was a pioneer in the application of chemical methods to the study of nutrition. In 1847, he isolated inosinate from beef stock, based on which he founded a company that produced bouillon cubes, later trademarked as Oxo. It is said that von Liebig remarked that hydrolyzed protein tastes good and meaty.

Walton, Izaak (1593–1683) author of *The Compleat Angler*, first published in 1653.

wasabi Japanese horseradish (*Wasabia japonica*).

Worcestershire sauce a type of fermented anchovy sauce, related to the classical Roman fish sauce *garum*. It can be made from anchovies, wine vinegar, molasses, salt, sugar, tamarind paste, soy sauce, cloves, lemons, pickles, pepper, onions, and garlic. In Western cuisine, the condiment most widely used to enhance umami is Worcestershire sauce.

xanthosine-5'-monophosphate (XMP) nucleotide derived from the nucleic acid xanthosinic acid. Interacts synergistically with glutamate to enhance umami taste; found, for example, in mushrooms.

xian-wei Chinese expression for deliciousness; has almost the same meaning as umami in Japanese.

yakiboshi see *niboshi*.

yeast extract hydrolyzed protein, different from nutritional yeast; produced by the autolysis of yeast cells. Marmite and Bovril are made from yeast extract.

yeast flakes see nutritional yeast.

yuba the skin that forms on heated soybean milk.

yu-lu Chinese fermented fish sauce.

yuzu small Japanese lemon (*Citrus junus*); used, for example, to make the dressing *ponzu*, which is a mixture of soy sauce, dashi, and *yuzu* juice, possibly also with a dash of mirin.

Index

A
Abbé M'Quin 159
acetic acid 26, 239
acetylsalicylic acid 32
acidity 5, 30, 196
adenosine-5'-monophosphate 35, 230. *See also* adenylate, AMP
adenosine-5'-triphosphate 11, 91, 239. *See also* ATP
adenylate 239. *See also* AMP
~ and umami 11, 35, 54
~ in fish and shellfish 35, 52, 69
~ in scallops 60, 163
~ in shrimp 53
~ in tomatoes 126–127
~ molecular structure 231
adenylic acid 11, 54, 239
aftertaste 6, 8, 16, 56, 60, 67
aging 208, 213
~ cheese 146, 148, 229
~ meat 138
~ seaweed 68
~ soy sauce 112–113
Ajinomoto 6, 20, 27–28, 30, 34, 93, 165, 239
alanine 82, 143, 239
~ in dashi and soups 161, 225
~ taste 39, 68, 160, 222
Alaska crab 226
alcohol 112, 127, 159, 199, 203. *See also* sugar, alcohol
algae 12, 23, 47, 66, 68, 125. *See also* seaweed
alkaloid 5, 239
allec 80, 239
allergy 33
allostery 221, 239
amino acids 239
~ and nutrition 5
~ chirality 10, 17, 29, 219

~ essential 10, 24, 106
~ free 10, 65, 137, 155, 199
~ from cooking 140
~ from fermentation 80–85, 96–97, 203
~ from hydrolysis 28–29, 143, 169–171
~ in cheese 119, 148
~ in dashi and soups 160–163, 225
~ in fish 52, 94, 97
~ in green tea 134
~ in konbu 68
~ in meat 143, 159
~ in milk 35
~ in miso 114
~ in *nattō* 120
~ in *niboshi* 91
~ in proteins 9–11, 12, 223
~ in rice, sake, and mirin 197, 200
~ in sea urchins, shrimp, and crab 69
~ in soy sauce 113
~ in yeast 172
~ neurotransmitters 17
~ receptor binding 219
~ salts 10
~ taste 10, 17, 18, 222
ammonia 98, 118
ammonium 10, 24
amniotic fluid 35
AMP 35, 62–63. *See also* adenylate
~ in foodstuff 230–231
~ in tomatoes 127
~ molecular structure 231
anchovy 46, 56, 63, 69, 80, 81–82, 91, 100–101, 174–175, 176–177, 179, 180, 191, 194, 226
~ paste 63, 81, 84, 175, 178, 230
ankimo 57, 239
ao-nori 125, 239
Apicius, Marcus Gavius 81, 82, 239

appetite 5, 207, 209–210
apple 63, 227
arabushi 90, 93–94. *See also katsuobushi*
arginine 240
~ taste 60, 69, 222
Aristotle 1, 14
aroma 6
arrowroot 124
artichokes 176
asparagine 222
asparagus 63, 105, 116, 164, 184, 198, 226, 230
aspartame 17, 18, 240
aspartate 240
~ and umami 10
~ in dashi and soups 161
~ molecular structure 222
~ receptor binding 219
~ taste 10, 162, 222
aspartic acid 10, 160, 240
~ in dashi and soups 225
~ taste 222
Aspergillus
~ *glaucus* 94, 240
~ *oryzae* 113, 114, 203, 240
~ *sojae* 113, 240
astringency 1, 7, 44, 240
ATP 11, 91. *See also* adenosine triphosphate
Australopithecines 141
autolysis 169, 172, 240
avgotaraho 98
avocado 227
azuki 202, 240

B
bacalao 240
Bacillus subtilis nattō 120, 240
bacon 145, 149, 151, 179, 191, 194
bacteria 29, 80, 120, 141–145

~ lactic acid 88, 96, 97, 106, 113, 114, 119, 148
bagna càuda 175–176, 179, 240
bakasang 81, 240
baking powder 33
Bang, Ivar 243
barley 106, 114–115, 119, 156
Bass Brewery 171
Bayonne ham 144, 240
beef 63, 138, 146, 162, 182, 224, 226, 231
~ corned 144–145, 241
~ cured 144–145
~ *estofado* 188–189
~ extract 170–171
~ jerky 143
~ maturation 143
~ patties 181–183
~ sausages 145
~ soup 34, 157, 162
~ tea 159
beer 28, 106, 171, 172, 199, 200–201
bell pepper 176, 226
Bellini cocktail 151
beriberi 171
biga 192, 240
biltong 143, 240
biscuits 149, 150
bisque 69, 74, 240
bitter taste 1–3, 5, 7, 16–20, 20, 35, 38, 50, 57, 90, 145, 149
~ amino acids 10, 69, 162, 222
~ and sweet 16–17
~ and umami 56, 127, 134, 168
~ in dashi 44–46
black bean paste 122
black pepper 7
blood pressure 55, 210
Bloody Caesar 127, 177
Bloody Mary 127, 177, 240
Blumenthal, Heston 126
bocconcini 146

Bolognese sauce 127, 240
bonito 43, 84, 90–91, 92–95, 156, 226, 230
botargo 98, *240*
bottarga. See *botargo*
bouillabaisse 69, *240*
bouillon 20, 156, 157, 159, 161, 164, 178, *240*. See also stock; soup
~ cubes 165
Bovril 170–171, *240*
brain 1, 6, 8, 16–17, 24–25, 33, 38–39, 42–43, 72, 217–218
~ evolution 138, 140–141
~ -mGluR4 217
~ scans 43
~ stomach 209
braising 201, *240*
brandade 190–191, *241*
brewing 65
~ beer 170–171, 199, 200
~ rice 106
~ soy sauce 112, 113
Brillat-Savarin, Jean Anthelme 154, 158–159, 166, 206, *241*
brioche 203–205, *241*
broccoli 226
brunoise 58, *241*
buckwheat 115
Buddhism 65, 66, 110, 122, 124, 173, 194
budu 81, *241*
butàriga 98
butter 146

C

cabbage 50, 82, 105, 106, 145, 182, 226
cabrales 227
caffeine 56, *241*
calcium 10, 24, 44
~ channels 6
~ inosinate 168
~ lactate 149
~ sulfate 119
calories 4–5, 36, 60, 140–141, 191, 210
Camembert 227
camphor 7

candy 202
capers 106
capsaicin 7, *241*
Car4 19
carbohydrates 4, 5, 15, 23, 27, 140, 158, 159
carbon dioxide 19, 96
cardiovascular disease 214
Carême, Marie-Antoine xvi, *241*
carpaccio 151, *241*
Carpaccio, Vittore 151
carrot 63, 155, 167, 200, 224, 226
casein 148, 169, *241*
cassoulet 182, 186–187, 197, *241*
catalyst 169, *241*
catfish 74
Cattle Market, The 101
cauliflower 50, 164, 167, 176, 226
caviar 60, 63, 98, 226
celery 155, 176, 190
cell
~ membrane 16, 24, 218–220, *241*
~ nerve/sensory 1, 6–7, 17, 25, 38, 209
~ presynaptic 16–17, 249
~ taste 6–7, 15–19, 217–221
~ wall 169, 172
~ yeast 110, 172
Cenovis 171, *241*
cereals 28, 106, 114, 115, 124, 168
cha-kaiseki 125, *244*
cheddar 148, 191, 227, 229
cheese 28, 63, 106, 118–119, 130, 142, 146–149, 224, 227. See also Parmigiano-Reggiano; cheddar; Parmesan
~ aging 146, 148–149, 229
~ and umami 4, 13, 142, 146, 179
~ blue 60, 63, 146–147, 179, 180, 196, 203, 227
~ hard 119, 146, 148

~ production 146–149, 169
chemesthesis 5–7, *241*
Chevrier, Canon 159
chewiness 6
chewing 6, 27, 140–141, 210
chicken 47–51, 49, 63, 138, 143, 159, 162, 224, 226, 231
~ bouillon 156–157, 161–162, 178, 197–199, 225
~ Marengo 184–185
chimpanzees 35, 140
Chinese
~ cabbage 226
~ cheese 120
~ radish. See daikon
~ restaurant syndrome 32–34, 33–35, *241*
chinuki katsuobushi 90, *245*
chirality and taste 10, 17, 29
chocolate 60, 203
chorizo 146, *241*
Cipriani, Giuseppe 151
clam 127
~ -bake 69, 74–75, 78
cod 57, 190–191, 224, 226, 230
~ roe 164
codability 20
coffee 56
confit 186, *241*
Confucius 136
consommé 45, 47, 156
cooking xvi
~ grandmother's 180
~ in history 140–141, 158–159
~ slow 182
corn 12, 28, 74, 164, 169, 173, 224, 226
crab 58, 63, 69, 74, 76–77, 226, 230
cranberry 182
cranial nerves 16, 209, *241*. See also vagus nerve
crayfish 184
crème fraîche 20, 146
crouton 76, 128, *242*

crunchiness 6, 197
cucumber 116, 182, 200, 226
cuisine. See also New Nordic Cuisine
~ Asian 4, 60
~ Japanese 2, 23, 36, 43, 114, 122–125, 156–157, 202, 214
~ Western 2, 36, 44, 60, 125, 157, 202
culinary arts 10, 36, 38, 42, 65, 66, 79, 81, 140–141, 213
curing 142
~ and umami 10, 137, 138
~ fish 65, 88, 96, 98
~ meat 143–146, 229
~ seaweed 68
cysteine 222

D

daikon 116, 125, *242*
dairy products 137, 142, 146–149, 173
~ glutamate 168, 227
Danablu 146
Danish blue 146, 196, 227
Darwin, Charles 140
dashi 12, 35, 39, 68, 156, *242*. See also konbu; dulse
~ amino acids 160–161, 225
~ and MSG 38
~ and umami 2, 20, 23–26, 34, 36, 43
~ bar 160
~ dulse 47–51, 225
~ *ichiban* 45
~ in soups 39
~ instant 47
~ Japanese food culture 43, 122, 156
~ konbu 46, 67, 160–161, 225
~ *niban* 45
~ *niboshi* 46
~ Nordic 47–53
~ preparation 43–52
~ *shōjin* 46, 47, 66, 125, 156, 194

256   Index

~ taste 25
dehydration 47, 53, 68, 92, 142, 144, 165, 223. *See also* drying
Dejima 113
deliciousness 8, 25, 36, 48, 79, 122, 124, 168, 213–214
deoxyribonucleic acid. *See* DNA
desserts 202–205
diabetes 214
digestion 5, 8, 140, 171, 209–210
DNA 11, *242*
donko 111, *242*
dorade 91
douchi 122, 228, *242*
dressing 55, 57, 80, 115, 146, 173, 175, 194
drying 10, 142. *See also* dehydration
~ and umami 65, 126, 137, 208, 229
~ fish and shellfish 79, 92–93
~ fungi 110
~ ham 229
~ meat 138
~ yeast 172
duck 57, 138, 162, 224, 226
dulse 68, 149–152, *242*
~ dashi 47–51, 225
Dunand 184

E
eggplants 115, 190, 200
eggs 63, 69, 74, 120, 142, 145, 151, 156, 178, 182, 184, 194, 224, 226
Eklund-Jonsson, Charlotte 119
El Bulli 48
elderly 207, 210–211
Emmental 227
ENaC 19
enlightened kitchen 124–125, 214. *See also shōjin ryōri*
enokitake 111, 230, *242*
enzyme 12, *242*
~ for cheese production 119

~ in fermentation 80–82, 85–86, 88, 96–98, 114, 120, 199, 203
~ in mushrooms 111
~ in yeast 110, 172
~ proteolytic 113, 169, 209, 249
epithelia 16, 19
estofado 182, 188–189, *242*
evolution 4, 19, 26, 36, 137–138, 140–141, 209

F
fast food 174, 191
fat 4, 57, 113, 143, 144, 148, 151
~ receptor 4
~ reduction 36, 168, 207, 214
~ taste 2, 5, 44, 159, 179, 191
Fat Duck, The 48, 126
fermentation 10, *242*. *See also* enzyme in fermentation
~ and umami 65, 105, 106
~ by *kōji* 113–115, 199
~ glutamate contents 228
~ of fish 23, 79–89, 96–98, 228
~ of *katsuobushi* 90, 93
~ of meat 142, 145
~ of milk 12
~ of rice 199–200, 228
~ of shellfish 87
~ of soybeans 111–115, 118–122, 228
~ of starch 28
~ of tea 54
fifth taste 3, 23, 36, 38, 156, 177
Fischer, Emil *243*
fish 46, 65, 91, 197, 224, 226. *See also* roe
~ and umami 4, 12, 52–53, 69, 84, 126
~ fermented 79–85, 88, 96–99, 114, 122
~ frozen 92
~ innards 80–82, 85, 92, 96–97

~ killing 72–73
~ liver 56–58
~ marinated 200
~ nucleotides 35, 54, 230
~ smoked 44, 86, 90, 92–93
~ soup 155–156, 162
fish sauce 6, 43, 63, 79–86, 92, 111, 148, 169, 174, 176, 228, 242. *See also garum*
flatfish 179
flavor 5–7, 60, *242*
~ definition 5
~ enhancer 27–28
floral taste 50
flying fish 91, 96
foie gras 120, 137, 138
~ from the sea 57
folic acid 171, *242*. *See also* vitamin B9
food additive 27, 33–34, 60, 167–169, 208
fowl 196, 231
French nouvelle cuisine 125
fricassee 184, *242*
fruits 5, 12, 141, 196, 214
~ and umami 105, 126, 168, 202, 227
fu 124, *242*
fundic glands 209, *242*
Funen Society of Serious Fisheaters 101
fungi 35, 46, 49, 51, 105, 110–111, 122, 124–125, 127, 164, 173, 174, 180, 207. *See also* mold; mushrooms; shiitake
~ glutamate 226
~ nucleotides 230
furu 119–120, *242*
fushi 90, *243*

G
game 114, 138, 143, 145, 182, 196
ganjang 112, *243*
garlic 6, 110, 116, 144, 175, 177, 179, 226
~ black 122

garon. *See garum*
garum 79–82, 164, 178, *243*
~ modern 85–87, 228
gelatination 140
Geoponica 85
gherkins 106
ginger 106, 125, 200
glucose 43, *243*
glutamate 24–25, *243*. *See also* MSG
~ and acidity 26, 30
~ and bitter 56
~ and glutamic acid 24–25
~ and salt 55, 56
~ and umami 3, 10, 24, 26–27, 30, 35, 38–39
~ as food additive 168–169, 208
~ as neurotransmitter 25, 217
~ bound 224
~ by fermentation 82–85, 96
~ chirality 29, 38, 219
~ daily intake 12
~ discovery 23, 25, 30
~ free 24, 223–229
~ in beans 122, 228
~ in beer 200
~ in bread 106
~ in cereals 106
~ in cheese 146–149, 227, 229
~ in corn 106
~ in dairy products 35, 146–149, 224, 226–227
~ in dashi and soups 23, 43–53, 155, 161, 225
~ in dulse 47–51
~ in eggs 151
~ in fast food 191
~ in fish and shellfish 69, 224, 226
~ in fish sauces 81–85, 228
~ in foodstuff 12, 62–63, 223–229
~ in fruit 105, 227
~ in fungi 110–111, 173, 226
~ in green tea 134, 226
~ in infant formula 36

Index 257

~ in konbu 30, 34, 68
~ in meat 9, 138, 143–146, 224, 226, 229, 231
~ in miso 114–115
~ in mother's milk 35–36
~ in nattō 120
~ in nuts and fruits 227
~ in potatoes 52, 105, 180, 197
~ in poultry 162, 224, 226, 231
~ in rice products 106, 199–200, 203
~ in roe 98
~ in sake 200, 228
~ in seaweeds 68, 227
~ in shellfish 87–88
~ in soy sauce 112–114, 228
~ in sushi 88
~ in tempeh 118–119
~ in the body 24–25
~ in the diet 12, 25, 33, 208, 211
~ in tofu 119–120
~ in tomatoes 44, 105, 126–127, 229
~ in vegetables 105, 116, 224, 226
~ in yeast 110, 169–173
~ ion 10–11, 26, 30
~ metabolism 25–26
~ production 27–29
~ receptors 25, 29, 38–39, 217–221
~ stability 44
~ synergy with nucleotides 34–35, 41–47, 60, 69, 72, 74, 155–156, 162, 218–221
~ taste 3, 13, 25–26, 29, 222
~ taste synergies 55, 218
~ taste threshold 13, 41–42, 223
glutamic acid 3, 9–12, 24–30, 33, 80, 110, 134, 143, 168, *243*. *See also* glutamate; MSG
~ acidity 30
~ by hydrolysis 27–28, 169

~ chirality 29
~ in dashi and soups 160–162, 225
~ in foodstuff 12, 24
~ in konbu 68
~ in milk 35
~ in proteins 106
~ stability 44
~ taste 10, 26, 222
glutamine 222
glutathione 6, 137, *243*
gluten 9, 28, *243*
glycine 52, 60, 144, *243*
~ taste 222
GMP. *See also* guanylate
~ as food additive 35
~ in foodstuff 62–63, 230–231
~ molecular structure 231
~ taste threshold 223
gohō 124
goma dōfu 124
gomi 124
Gorgonzola 146
goshoku 124
Gouda 227
gourmet ape 5
G-protein 16, 17, 218, 220
~ -coupled receptor 16, 17, *243*
grandmother's cooking 180
grapefruit 227
grapes 28, 227
gravlax 97
gravy 179, 180, 196, 197
grilling 53, 91
Gruyère de Comté 227
guanosine-5′-monophosphate 34. *See also* guanylate; GMP
guanylate 11, 34–35, 36, *243*. *See also* GMP
~ in foodstuff 230–231
~ in fungi 110, 180, 184, 230
~ in meat 143, 231
~ in nori 68, 231
~ in potatoes 197, 230
~ in shiitake 46, 110–111, 156, 230
~ in tomatoes 126, 230

~ molecular structure 231
~ receptor binding 221
~ synergistic umami 41, 49, 51, 105, 180, 219–221
guanylic acid 11, 34, *243*
gunkan-maki 88–89, *243*
gustducin 17, 220, *243*
gyokuro 134, *243*

H
habilines 141
hákarl 96–98
ham 167, 229
~ air-dried 4, 9, 63, 108, 142, 143–144, 179, 192, 226
~ smoked 49–51
hamburger 151, 174, 191
Han Tombs 111
Harry's
~ Bar 151
~ crème 151–152
harsh taste 1
Heinz ketchup 174
hemp seeds 125
herring 82, 96, 151, 226
Hidaka-konbu 67. *See also* konbu
hidoku 92
hishio 112, *243*
histidine 94, *243*
~ taste 222
hōjicha 134, *243*
Hokkaido 66
homeostasis 209, *243*
Homo
~ *erectus* 141
~ *sapiens* 5, 138, 140–141, 211
Hon-dashi 47, 90, 93
honey 13, 20, 79
HP sauce 182–184, *243*
hydrogen ions 16, 19
hydrolysis *243*
~ in vegetables 169
~ in yeast 110, 169
~ MSG production 27–28
~ of proteins 169

hydrolyzed protein 168–169, 244

I
ika no shiokara 82
Ikeda, Kikunae 22–27, 38, 40, 64, 65, 217, 244
ikijime 72, 244
immune system 12, 209, 210
IMP. *See also* inosinate
~ as food additive 35
~ in foodstuff 62–63, 230–231
~ in katsuobushi 34
~ in meat 231
~ molecular structure 231
~ taste synergy 42–43
~ taste threshold 223
innards 80–82, 85, 96–97, 137
inosinate 11, 244. *See also* IMP
~ as food additive 168
~ in bacon 149
~ in fish 44, 52, 69, 72, 175
~ in fish sauce 79, 84
~ in katsuobushi 34, 90, 92, 94
~ in mackerel 127
~ in meat 34, 35, 49, 138, 143, 145, 162, 180
~ in niboshi 46, 91
~ in oysters 100
~ in roe 98
~ in shrimp 53
~ molecular structure 231
~ receptor binding 219–221
~ synergistic umami 34, 35, 39, 41, 42
~ taste threshold 13, 41
inosine-5′-monophosphate. *See also* inosinate; IMP
inosinic acid 11, 34, 244
intestines 12, 24, 80, 209
ion channel 7, 16, 17, 19, 218, 244

irori 43, 84, *244*
irritant 6, *244*
ishiri 81–82, 84, *244*
ishiru. See *ishiri*
isoleucine 222
isothiocyanate 7, *244*

J

jamón serrano 143–144, *244*
jerky 143, *244*
jiàng 111, 112, 114, *244*
Johnston, John Lawson 170

K

kabayaki 123, *244*
kaiseki 124 125, *244*
karebushi 90, 93, 94. See also *katsuobushi*
katsuo 43, 84, 90, 91–92, *244*
Katsuo Gijutsu Kenkyujo 92
katsuobushi 88, 90, 230, *245*. See also *arabushi*; *karebushi*
~ and umami 34, 63, 84
~ in dashi 23, 34, 43–45, 47, 160–161
~ production 92–95
~ taste 94
ketchup 55, 63, 80, 126, 174, 191, *245*
kimchi 82, 105, *245*
kitchen as a lab 48–51
kiwifruit 227
Knorr, Carl Heinrich Theodor 164, 165, *245*
kobujime *245*
Kodama, Shintaro 34, *245*
koe-chiap 174, *245*
kōji 113, 114, 199, *245*
kokumi 6, 122, 137, *245*
kombucha 54, *245*
konara 93
konbu 30, *245*. See also *Hidaka-konbu*; *ma-konbu*; *oboro-konbu*; *Rausu-konbu*; *Rishiri-konbu*; *tororo-konbu*
~ amino acids 68

~ and umami 34, 63, 68, 156
~ as foodstuff 65–68
~ cultivation 68
~ glutamate 12, 227
~ harvest 66
~ in dashi 23, 39, 43–47, 50, 90, 160, 225
~ road 66
Kuninaka, Akira 34, 39, *245*
kunugi 93
kuragakoi 68, *245*
Kuriwaki, Mio 92
kusaya 96, *245*
kuzu 124
Kwok, Robert Ho Man 32 34
Kyoto 66, 160

L

lactic acid. See bacteria, lactic acid
lactisole 219, *246*
lactose 36, *246*
lamb 138, 159, 184, 226
Laminariales 68
laver 68. See also *Porphyra*
Lea & Perrins 176
lecithin 151, *246*
lemon 13, 20, 98, 125, 177, 179, 200
lenthionine 111, *246*
lentils 105
Lentinus edodes *246*
lettuce 194, 226
leucine 143, 148, *246*
~ taste 222
lipid 17, *246*
liquamen 79, 80, *246*. See also *garum*
liquid aminos 112
liver 24, 25, 56–57, 137
lobster 54, 63, 70, 74, 78, 226, 230
lotus root 123–125, 226
lutefisk 96
lysine 148, 162, *246*
~ taste 222

M

maccha 134, 177, *246*
mackerel 63, 69, 74, 80, 85–87, 90, 91, 96, 127, 164, 224, 226, 230
Maggi
~ cube 164–165
~ Julius 165, *246*
~ sauce 165, 169
magnesium 10, 24, 44
~ chloride 119
magurobushi 90, *246*
Maillard reactions 122, 162, 182, *246*
maize 106
ma-konbu 67. See also konbu
Makurazaki 90
mandarin peel 125
Manganji pepper 123, *246*
mannitol 23, 30, 68, *246*
Marengo 184
marinade 88, 177, 178–179, 200
marinating 10, 67, 97, 110, 128, 143, 174, 178, 200, 223
Marmite 63, 169, 170–171, *246*
marrow 116, 226
Matsuhisa, Nobu 39
matsutake 110, 111, 230, *246*
maturing
~ cheese. See ageing cheese
~ fruits and vegetables 105
mayonnaise 142, 151, 191
McGee, Harold 216, *246*
meat 65, 122, 137, 140–141, 151, 178, 182, 191, 200
~ and umami 2, 127, 137–138, 146, 160–161, 170, 181, 184, 231
~ extracts 158–159
~ glutamate 12, 161–162, 224, 226
~ in soup 4, 44, 155, 157–162, 180, 220
~ nucleotides 35, 231
~ preservation 142–146

~ smoked 142, 144–145
meaty taste 26, 149, 165
medisterpølse 180–181, *246*
Meiji era 84
meligarum 79. See also *garum*
melon 20, 200
membrane 7
~ cell 15–16
~ mucous 6, 24
~ potential 16
~ protein 16–19, 217–221
menthol 7
metabolism 5, *246*
~ of glutamate 24 25
metallic taste 51, *246*
methionine *246*
~ taste 69, 162, 222
mGluR1 39, *246*
mGluR4 217, 219, *246*
mice 18, 19, 33, 38, 219
Micrococcus glutamicus 29, *247*
microvilli 15, *247*
milk 35–36, 56, 142, 146, 175, 179, 224, 226
~ cow's 12, 35, 63, 142, 146–149, 224, 226
~ fermented 12, 28, 119–120, 137, 142, 146–149, 169
~ human breast 12, 35–36, 211, 224, 226, 231
~ sheep's and goat's 146, 226, 227
~ soy 106, 118–119, 125
millet 115
mirin 123, 125, 200, 203, *247*
miso 28, 63, 110, 111–112, 114, 124, 173, 177, 178, *247*
~ glutamate 228
~ production 114–115, 203
~ soup 27, 33, 38, 45–47, 91, 116, 156
~ varieties 115
~ -zuke 116
mojama 98, *247*

molasses 28, 172, 177, 247
mold 90, 94, 118, 142–143, 144. See also fermentation
molecular gastronomy 38, 39, 47, 126
mollusks 81
monkfish 57, 58
monosodium aspartate. See aspartate
monosodium glutamate. See glutamate
morel 110–111, 164, 230
moromi 113, 247
mouthcoating 6, 151
mouthfeel 5–6, 125, 151, 168, 247
MSA 10, 222, 247. See also aspartate
MSG 3, 10, 30, 247. See also glutamate
~ and umami 36
~ as food additive 25, 27–28, 84, 87, 151, 167–169, 207–208
~ 'Chinese restaurant syndrome' 32, 34
~ chirality 29
~ daily consumption 28
~ discovery 23–24
~ in foodstuff 62–63, 226
~ in tomatoes 127
~ molecular structure 29
~ production 27–30
~ purity 29–30
~ safety issues 168, 208
~ stability 30
~ taste 3, 13, 20, 38, 43, 55–56
~ taste threshold 13, 42, 55, 223
mullet 98, 132
muria 80, 247
mushrooms 47, 49–51, 52, 63, 88, 105, 110–111, 118, 138–139, 164, 178, 184, 191, 226, 230. See also shiitake; oyster; porcini

mussels 69, 74, 76, 226
mustard 7
mycelium 113, 118, 247

N
nabe 156, 247
nam-pa 81, 247
nam-pla 81, 247
nama-zushi 88
Napoleon 170, 184
nare-zushi 88, 96. See also sushi
nasu dengaku 115, 247
nattō 111, 118, 120–121, 228, 247
néré 122
nerve. See also cell, nerve/sensory; cranial nerves
~ fiber 16
~ impulse/signal 1, 17, 25, 39, 218
neurotransmitter 17, 25, 217, 248
New Nordic Cuisine 47, 48, 248
ngan-pya-ye 81, 248
ngapi 87, 248
niboshi 46, 91, 226, 230, 248
nimono 156, 248
Ninomiya, Kumiko 39, 92, 126
Nishiki food market 160
nojime 72, 248
noma 48
noodles 35, 156, 191, 203
Nordic Food Lab 48–51
nori 68, 88, 227, 231, 248. See also *Porphyra*
nucleic acid 11–12, 34–35, 54, 248
~ and taste 155
nucleoside 248
nucleotides 248. See also ribonucleotide; AMP; GMP; IMP
~ and taste 11–12, 41–43
~ and umami 3, 39, 168, 217–221
~ discoveries 34–35

~ free 11–12, 65
~ in foodstuff 62–63, 230–231
~ molecular structure 231
~ stability 35
~ taste threshold 13
nuoc mam tom cha 81, 84, 248
nutrition 4–5, 10–11, 27, 36, 43, 60, 65, 88, 116, 140–141, 142, 155, 169–171, 200, 209–210
nutritional yeast 149, 172–175, 179, 203, 248
nuts 106, 143, 227

O
oat 119
obesity 209, 214
oboro-konbu 67. See also konbu
odor 6
oenogarum 79
onion 97, 106, 177, 181, 182, 190, 226
Onozaki, Yoshitaka 124
orbitofrontal cortex 43
orthonasal 6, 248
Osaka 66, 67
osmazome 154, 158–159, 248
osso buco 182, 248
Ostwald, Friedrich Wilhelm 27, 248
oxtails 201
oxygarum 79. See also garum
oyster 4, 20, 63, 69, 87, 100–101, 116, 175, 226
~ mushrooms 63, 111, 230
~ sauce 81, 87–88

P
palatability 8–9, 60, 122, 142, 171, 194, 207
~ and umami 123, 211, 214
~ glutamate 25, 55, 208
Palmaria palmata 68, 248. See also dulse
pancetta 194, 248

panko 115, 174–176, 248
papillae 15
Parkia biglobosa 122, 228
Parma ham 144
Parmesan 63, 130, 148–149, 150, 156, 173, 177, 179, 191, 192, 194
Parmigiano-Reggiano 108, 146, 148–149, 150, 194, 198, 249. See also Parmesan
~ and glutamate 148, 224, 227
pasta 98, 127, 130, 149, 178, 179, 191
pastrami 144
pata negra 143, 249
pâté 178, 196
patis 81, 249
pear 227
peas 12, 63, 105, 145, 163, 164, 190–191, 224, 226, 230
Penicillium roqueforti 146, 249
peppermint 7
pepperoni 146, 249
peptide bond 9
peptides 10, 143, 148, 155, 169, 249. See also polypeptides
~ taste 6, 54
Pepys, Samuel 98
perch 74
pesto 149, 178
phenylalanine 148, 249
~ taste 222
Physiology of Taste, The 158–159
pickling 105, 106, 249
pigs 143, 159
pike 100–101
piperine 7, 249
pizza 146, 179, 191, 192
PKD2L1 19
plants and umami 35, 105, 168, 169, 173
Pliny the Elder 79
polysaccharides 50
ponzu 57, 110
popcorn 172
porcini mushrooms 49, 51, 111, 177, 230

pork 47, 63, 138, 143, 145, 162, 167, 180–181, 224, 226, 231
Porphyra 68, 227, 231, 249. *See also* laver; nori
porridge 159
potassium 10, 24, 32
~ and saltiness 5
~ inosinate 168
~ ions 16, 19
potato 63, 74, 97, 164, 181–182, 191, 224, 226, 230
~ and umami 12, 86, 105, 180, 197
~ water 52–53, 197
poultry 56–57, 138, 156–157, 162, 179, 182, 184, 224, 226, 231
preserving 142
presynaptic cell 16–17, 249
primates 35, 141
proline 68, 249
~ taste 222
Promite 171, 249
prosciutto 144, 249
protease. *See* proteolytic enzyme
proteins 249. *See also* hydrolyzed protein; receptor
~ and amino acids 9
~ breakdown of 28, 65, 80, 92, 97, 110, 113, 119, 120, 143, 148, 157, 172, 199, 209
~ coagulation 119
~ in foodstuff 12, 24, 27, 36, 106, 137
~ in nutrition 4–5, 10, 209
~ in the diet 33, 158
~ taste 10, 15, 18
proteolytic enzyme 80, 209, 249
pungent flavor 1, 87, 98, 120, 122

Q
quail 196
quinine 2, 249

R
Rabelais 159
radicchio 20
ragout 182, 249
rakfisk 96–97, 98, 249
ratatouille 190, 249
rats 38, 218, 219
Rausu-konbu 67, 225. *See also* konbu
receptor 249. *See also* G-protein-coupled receptor
~ cells 15–17, 218
~ chirality 219
~ combinations 18
~ fat 4, 5
~ function 16–19, 209, 218–221
~ glutamate 11, 25, 29, 209, 217–221
~ olfactory 6
~ pain, touch, temperature 5
~ sweet, sour, salt, bitter 3, 15–19
~ taste 1–8, 60
~ umami 3, 27–28, 38–39, 60, 92, 209, 217–221
red currants 20, 182, 196
rennet 119, 148, 249
retronasal 6, 250
Rhizopas
~ *oligosporus* 118, 250
~ *oryzae* 118
ribonucleotide 3, 11, 34–35, 39, 41, 43, 47, 54, 60, 69, 84, 88, 90, 91, 110, 126, 155, 194, 217–220. *See also* nucleotide
~ in foodstuff 230–231
rice 84, 88, 94, 112, 114, 115, 120, 156
~ and umami 106, 119, 149, 197
~ fermentation 199–200, 203, 228. *See also* mirin; sake: rice vinegar
~ vinegar 67, 88, 106
ricotta 146
rigor mortis 72, 244

rikakuru 84, 250
ripening 137
~ cheese 148
~ meat 143, 144
~ tomatoes 229
Rishiri-konbu 67, 90, 225. *See also* konbu
risotto 250
Ritthausen, Karl Heinrich Leopold 243
RNA 11, 250
roe 69, 88, 98, 164, 226
Roman cuisine 79–83, 85
Roquefort 146, 196, 203, 204, 227
rouille 69, 250
rye 106, 107, 115, 144

S
Saccharina japonica 12, 23, 68, 227, 250
saccharine 16, 250
Saccharomyces cerevisiae 172, 250
Sakai 67
sake 54, 57, 106, 112, 116, 197, 228
~ and umami 199–200
~ *kasu* 200, 250
SAKI Bar & Food Emporium 124
salad 126, 146, 173, 178, 179, 194, 214
~ Caesar 175, 194
salami 146
saliva 7, 15, 42, 210
salmon 82, 87, 224, 226, 230
~ roe 88, 98, 226
salsa 126
~ *verde* 58, 250
salsiccia secca 146, 250
salt. *See also* sodium
~ reduction 55, 210, 214
salting 65, 142
~ and umami 137
~ cheese 148
~ fish and shellfish 81–82
~ meat 138, 144–146, 229
salty taste 1–3, 5, 7, 13, 16, 18–20, 26–27, 51, 79, 112, 170, 174

~ and umami 20, 30, 55, 56, 84, 127, 145
sandwich 151, 181
San Gabriel, Ana 92
sanshō 123, 125, 250
sardine 46, 69, 82, 91, 226, 230
sashimi 72, 125, 250
sauce 35, 69, 78, 79, 100, 122, 127, 128, 146, 149, 175, 178–179, 180. *See also* fish sauce; HP sauce; oyster; soy sauce; Tabasco; tomato; Worcestershire
sauerkraut 105
sausages 35, 145, 145–146, 167, 180
sautéing 159, 182, 250
savory taste 5, 36, 69, 72, 112, 116, 119, 167, 179, 180, 182, 191, 196, 210, 213–214
scallop 6, 54, 57, 60, 63, 69, 163, 226, 230
sea
~ asparagus (beans) 20
~ lettuce 125, 173, 250
~ salt 30, 51, 68, 98
~ urchin 69, 226, 230
seaweed 23, 63, 74, 124, 231. *See also* algae; dulse; konbu; nori; sea lettuce; *wakame*; winged kelp
~ and umami 65–68, 122
~ glutamate 12, 25, 27
~ in clambake 78
~ in dashi 44–47, 49–51, 52, 225
seitan 173, 250
sencha 134, 250
senji 84, 250
sensory science 5–7, 250
serine 222
Shackleton, Ernest Henry 170
shark 97–98, 162
shellfish 52, 65, 69, 74, 174, 197, 226, 230
~ and umami 54, 69, 126, 184

~ fermented 81, 87, 114
~ nucleotides 35, 88
~ soup 156, 162, 178
shichimi 123, 125, 250
shiitake 36, 63, 110–111, 125, 178, 226, 230, 250
~ and umami 105, 110–111, 164, 177, 191
~ in dashi 46–47, 110, 156
~ taste 110
shiokara 82, 250
shōchū 203, 250
shōjin
~ dashi 46–47, 66
~ *ryōri* 65, 112, 122–125, 173, 250
shōyu 112, 124, 156, *251*. See also soy sauce
~ production 112–114, 203
shrimp 54, 63, 69, 226, 230
~ heads 52, 53, 127, 175
~ paste 87
simmering 44, 137, 145, 156, 159, 162, 194
~ and umami 180, 182
smell 3, 5–6
smoking 10, 53, 65, 90, 92, 137, 138, 142
smorgasbord 181
smørrebrød 181
sodium 10, 11
~ and saltiness 5, 19, 24, 32, 51, 55
~ and taste 55
~ aspartate. *See* aspartate; MSA
~ channels and pumps 16, 19
~ chloride 24, *251*
~ glutamate. *See* glutamate; MSG
~ ions 11, 30, 35, 55
~ nitrite 145
sorbet 202, *251*
soufflé 79, 146, *251*
soumbala 122, 228, *251*
soup 4, 13, 23, 34, 35, 44, 69, 149, 155–165, 170, 178–179, 220, 225. *See also* bouillon; dashi; stock

~ amino acids 160
~ bird's nest 162
~ powder 47, 93, 165
~ shark's fin 162
soupe des primes 159
sourdough 106–107, 107
sour taste 1, 3, 11, 13, 16, 18–20, 25, 27, 35, 38, 197, 200
~ amino acids 222
Soutatu 160
soybeans 28, 81, 105, 111, 114–115, 118, 120, 122, 164, 169, 228
~ and umami 106, 226
soy production 112–113
soy products 106, 111, 112–115, 118–122, 125. *See also shōyu*; miso; tofu
soy sauce 32, 57, 84, 88, 106, 112–114, 178, 203, *251*. See also *shōyu*
~ and Japanese food culture 156
~ glutamate/umami 30, 63, 112, 182, 191, 197, 228
~ history 112–113
~ production 26, 28, 81
spaghetti alle vongole 127
spicy taste 1, 7, 26
spinach 105, 224, 226
spoilage 65, 142, 143
squid 54, 69, 82, 226, 230
starch 28, 88, 128, 140, 156, 158, 199, 200
stew 165, 167, 179, 182, 184, *251*
Stilton 146, 227
stock 39, 47, 52, 162, *251*
~ amino acids 160–161
~ and dashi 12, 20, 23, 43–46, 48–51, 66–67, 125, 127, 156, 225
~ fish and shellfish 162
~ meat 138, 144, 156, 162, 178, 180, 225
~ -pot 159
~ soup 155–165, 165, 225

~ vegetables 156, 180
stomach 24, 33, 92, 119, 141, 209
strawberries 227
Stroganoff *251*
succinic acid 54, 200, *251*
sucrose 30, *251*. *See also* sugar
sufu. *See furu*
sugar 2, 5, 28, 30, 52, 55, 88, 97, 105, 162, 168, 172, 174, 199, 200, 202, 203
~ alcohol 23, 68
~ kelp 47, 49–50, 68
~ reduction 202, 207, 214, 218
suimono 156, *251*
sulfurous taste 35, 50, 69, 222
sunflower seeds 106
surströmming 96, *251*
sushi 125, 197
~ and umami 68, 88
Suzuki, Saburosuke 27
Suzuki, Toshio 72
sweetener 16, 17–18, 60, 219
sweet taste 1–3, 5, 7, 13, 16–18, 20, 26, 36, 38
~ amino acids 10, 52, 68, 160, 162, 222, 225
~ and bitter 16–17
~ and umami 43, 55, 127, 149, 162, 202, 218–219
sweets 202
swordfish 98

T
T1R 17, 38, 218, 252
T1R1 18, 219–221
T1R1/T1R2 18
T1R1/T1R3 18, 38–39, 218–221
T1R2 18, 219
T1R3 18, 219–221
T2R 17–19, 252
Tabasco sauce 76, 127, 152, *251*
Tabella Cibaria 159
table salt 2, 24, 28, 30, 32–33, 34, 51, 55, 208, *251*

~ and umami 3, 26, 55, 56
~ taste threshold 13, 26
tamari 112, 173, *251*
tamarind 122, 177, *251*
tang 161–162, 225, *251*
tannic acid 251
tannin 7, *251*
tapas 98
taste *251*. *See also* receptor; *See also* bitter; sour; sweet; umami
~ basic 1–5, 7–8, 20, 23, 27, 38–39, 60, 124, 168, 174
~ buds 3, 4–6, 15–16, 19, 38, 209, 252
~ enhancer 3, 20, 27–29, 35, 60, 79, 84, 111, 168, 170, 174, 176, 252
~ floral 47
~ history of 1–3
~ in evolution 4–5
~ intensity 13, 30, 35, 41, 42
~ language 20, 36
~ map 7
~ metallic 51, 246
~ preference 35–36, 213
~ receptor cells 252
~ synthetic 60
~ thresholds 7, 13, 41–42, 55, 223
taste-mGluR4 38–39, 217–220, 252
Taste No. 5 177
tea 7, 44, 54, 105, 134, 202, 226. *See also gyokuro*; *hōjicha*; *kombucha*; *maccha*; *sencha*
~ and umami 134
~ ceremony 125
tempeh 118–120, 228, 252
tempura 156, 203, 252
teriyaki 203, 252
teuk trei 81, 252
texture 5–6, 72, 115, 118, 120, 127, 143, 171, 197, 252
theanine 54, 134, 252
thermal perception 7
third spice 3, 167, 252
This, Hervé 104, 252

threonine 222
thrush 196, 252
tofu 66, 67, 106, 111, 118–120, 125, 173, 252
~ stinky 119–120
tomatillo 105
tomato 55, 63, 105, 126–128, 156, 164, 174, 177, 182, 184, 190–193, 202
~ and umami 4, 44, 56, 126, 194
~ glutamate 12, 52, 126, 224, 226, 229
~ juice 55, 86, 127, 177, 202
~ nucleotides 126–127, 184, 230
~ pulp 126–127
~ sauce 127, 130, 164, 191–192
Tomimatsu, Tooru 92
tongue 1, 3–4, 6–8, 15, 38, 42, 209, 220
tororo-konbu 67. See also konbu
Tosa 90
Tosa, Youchi 90
trigeminal nerve. See cranial nerves
trimethylamine 60, 252
~ oxide 97, 98
truffles 110, 191, 198–199, 226
tryptophan 222
Tsuji, Shizuo 156
tsuyu 203, 252
tuna 44, 69, 80, 98, 226, 230

turkey 138
tyrosine 149, 222

U

umai 2, 23, 26
umami
~ and wellness 207–211
~ basal 41, 54, 62–63, 105, 155
~ definition 26, 252
~ demonstration 126
~ global presence 36
~ research 34, 38, 39
~ synergistic 11, 34, 41, 62–63, 105, 155, 220–221
~ twelve easy ways 178–179
Umami Burger 191
Umami Information Center 39, 92
Umami Mama 39
Undaria pinnatifida 227

V

vagus nerve 209, 252
valine 148, 222
veal 138, 144, 151, 152, 184
vegan 122, 170, 173, 177, 252
Vegemite 169, 171, 252
vegetables 74, 82, 106, 116, 120, 122, 174, 178–179, 184, 190, 194, 200, 214
~ and umami 2, 69, 105, 116, 162, 164, 167
~ glutamate 12, 169, 180, 182, 224, 226

~ in soup 4, 44, 155–156, 160, 180, 220
~ nucleotides 230
vegetarian 45, 46–47, 52, 65, 66, 88, 112, 122–125, 130, 149, 156, 164, 170, 173, 177, 179, 194
Venus flytrap 220–221
vinegar 2, 26, 33, 67, 79, 88, 96, 106, 118, 122, 143, 151, 174, 177, 208
viscosity 6
vitamin 116, 141, 252
~ B 118, 171, 172
~ B9 171. See also folic acid
~ B12 171
~ K 120
vodka 127
von Liebig, Justus 34, 171, 253

W

wakame 227
walnuts 20, 63, 106, 174, 227
Walton, Izaak 100–101, 253
wasabi 253
Watanabe, Ayako 124
water quality 44–45
wellness 207–211
wheat 106, 112–113, 115
~ beer 200–201
~ protein 9, 28, 124, 169, 173
wine 7, 68, 79, 101, 199, 203

winged kelp 49–51
Worcestershire sauce 63, 78, 80–81, 127, 151–152, 176–177, 178, 180, 182, 194, 197, 253
Wrangham, Richard 140–141

X

xanthosine-5'-monophosphate 35, 253. See also XMP
xian-wei 253
XMP. See xanthosine-5'-monophosphate

Y

Yaizu 90, 92–94
yakiboshi 91. See also *niboshi*
Yamaguchi, Shizuko 42
Yanagiya Honten 92
yeast 29, 34, 110, 113, 114, 169–171, 199, 200
~ baker's 106, 172
~ extract 6, 110, 168–172, 253
~ nutritional 149–150, 172–173, 179, 248
yogurt 28, 142, 146
yuba 125, 253
yu-lu 81, 253
yuzu 57, 125, 177, 253

Z

Zen 66, 122, 124

THE PEOPLE BEHIND THE BOOK

Ole G. Mouritsen is a scientist and professor of biophysics at the University of Southern Denmark. His research concentrates on basic science and its practical applications to biotechnology, biomedicine, and gastronomy. He has received a number of prestigious science and science communication prizes. In his spare time, he cooks and furthers his knowledge of all aspects of food.

Klavs Styrbæk is a celebrated and award-winning Danish chef and owner of Restaurant Kvægtorvet ("The Cattle Market") in Odense. His book *Mormor's mad* ("Grandmother's Food") was honored with a special jury prize at the Gourmand World Cookbook Awards in 2007.

Jonas Drotner Mouritsen is a graphic designer and owns the design company Chromascope (www.chromascope.com). His movie projects have won several international awards.

Mariela Johansen has Danish roots, lives in Canada, and holds an MA in humanities.